Born in Chamonix, France, Paul-Henri Nargeolet lived in Africa with his parents for 13 years. After a 22-year career in the National Marines, he was awarded the grade of Commander. He then joined the French Institute for Sea Exploration (IFREMER) where he specialized in deep-sea exploration in 1986. The following year, he led the first recovery mission on the Titanic. He was the Director of the Underwater Research Program for RMS Titanic, Inc. (USA). In that capacity, he supervised the recovery of some 5,000 artifacts from the site of the wreck. He was one of the five victims who died onboard the *Titan* submersible on June 18, 2023.

THE SECRETS OF THE TITANIC

PAUL-HENRI NARGEOLET

Co-authored by
JEAN NOËL MOURET

Translated by
LAURA HAYDON

One More Chapter
a division of HarperCollins*Publishers*
The News Building
1 London Bridge Street
London SE1 9GF

www.harpercollins.co.uk

This paperback edition 2024

24 25 26 27 28 LBC 6 5 4 3 2

First published in Great Britain in ebook format by
HarperCollins*Publishers* 2023

A catalogue record for this book
is available from the British Library

ISBN: 978-0-00-869408-1

Typeset in Palatino
by Palimpsest Book Production Ltd, Falkirk, Stirlingshire

Printed and bound in the United States

To my friend Captain Philippe Tailliez, father of all French Navy divers, who steered me towards diving as a teenager, and subsequently into underwater intervention. His motto: 'Enthusiasm is the only virtue.'

To my friend George Tulloch, with whom I spent much time at the Titanic site and who introduced me to his country, the United States of America. Without him, this entire story would probably never have occurred.

RMS *Titanic*: Key Information

Shipowner: White Star Line, Liverpool.
Co-owners: White Star Line (UK), International Mercantile
Marine Company (USA).
Home port: Liverpool.

Class: Olympic.
Sister ships: RMS *Olympic* (launched in 1911), RMS
Britannic (launched in 1914).
Architects: Thomas Andrews (1873–1912, perished with
the ship), Alexander Montgomery Carlisle (1854–1926).
Site: Harland & Wolff, Belfast, Northern Ireland.
Hull: Riveted steel, 2,000 plates measuring 3 x 2 metres
each, held together by roughly 3 million rivets.
Length: 269 metres.
Main beam (the widest point of the vessel): 28 metres.
Total height (including funnels): 53.10 metres.
Air draft: 45 metres.
Draft: 10.54 metres.
Power: Between 46,000 and 51,000 horsepower (maximum
power with steam turbine backup).
Cruising speed: 21 to 22 knots.
Maximum speed: 23 to 24 knots.
Displacement (weight of the volume of water displaced
by the vessel afloat): 5,310 tonnes.

Crew numbers: 885.
Maximum number of passengers: 2,471.

Construction start date: 21st March 1909.

Launch: 31st May 1911.

Sea trials: 2nd April 1912, 6am to 6pm.

Commissioning: 10th April 1912.

Maiden crossing and shipwreck:

Departure from Southampton (England): 10th April 1912, at 12.15pm.

Stopover in Cherbourg: 10th April 1912, from 6.35pm to 8.10pm.

Stopover in Queenstown (now Cobh, Republic of Ireland): 11th April 1912, from 11.30am to 1.30pm.

Collision with an iceberg: 14th April 1912, at 11.40pm.

Shipwreck: 15th April 1912, at 2.20am.

Arrival of the *Carpathia*: 15th April 1912, at 3.30am; first lifeboat rescued at 4.10am.

On board during the crossing: 889 crew members, 1,326 passengers.

Result: Between 1,491 and 1,513 fatalities, 705 survivors.

Discovery of the wreck: 1st September 1985.

First recovery of items: 25th July 1987.

Last survivor: Eliza Gladys 'Millvina' Dean, who died on 31st May 2009, aged 97.

NB: All times are local.

Tribute to Paul-Henri Nargeolet

On June 22, 2023, once the discovery of debris from the *Titan* submersible had been confirmed, the world's media headlined the disappearance of 'Mr. Titanic'. But for us, his children, and grandchildren, he was first and foremost Dad and Grandpa. A father and grandfather not quite like any other, he was determined to pass on his philosophy for life: first of all, *go all out for the things you love to do*. For him, doing things out of obligation was strictly forbidden. Work had to be at least a pleasure, preferably a passion. To live one's life consisted, according to him, in having goals and giving one's all to achieve them. The second prohibition was routine. He hated routine, to the point of doing everything he could to escape it: 'I make a habit of never having a habit', he often said.

Our grandparents lived in Chamonix, where our grandfather was a pharmacist. When our grandmother was seven months pregnant, she was at her hairdresser's when there was an explosion in the pharmacy on the opposite side of the road. She was so frightened that she gave birth prematurely. In 1946, a seven-month-old premature baby had very little chance of survival, but he pulled through. Our father often laughed about it, saying that having been born in an explosion, he could only become a clearance diver, specialising in explosives. And his death, too, was linked to an explosion-implosion – a whole existence written between two deflagrations.

As for his passion for the sea, he has lived it to the full, every day of his life since childhood. Perhaps it's because he was born under the water sign of Pisces. In any case, this mountain man became hooked on diving at the age of 8 or 9, when his parents were stationed in Morocco. After school, he would set off alone with his fins, mask and snorkel, sometimes risking his life – on more than one occasion, he had been swept away by the current and barely escaped – but he always came back.

Later, he chose to become a clearance diver in the French Navy, a high-risk career. From the training and his experiences, he acquired a 100% awareness in the importance of safety. He was known to be very protective of his crew members, concerned about their comfort and working conditions. During diving missions to the *Titanic,* he sometimes refused, as expedition director, to lower the submersible due to bad weather conditions. To those who argued that this was a waste of time and money, he objected, 'We're not in the business of putting people at risk just because we're in a hurry to finish the mission'. An attitude that didn't win him many friends, but as long as he saw the slightest danger, his position remained unchanged: 'We're not diving'. At the same time, he knew how to play down the situation without turning a blind eye to it. In a video, he recounts how, on a very stormy day, a member of crew on the accompanying boat became worried and asked him, 'could we sink?', to which he replied with a laugh, 'of course, we could sink!'. That's just like him, refusing unnecessary risks but accepting the inevitable dangers. In fact, he talked about them as little as possible to protect us. He sometimes mentioned consequences, but he never went so far as to say, 'if this happens, I'll die tomorrow'.

His all-consuming passion for the sea went through three stages: first diving, practiced since childhood and adolescence, then underwater research with the French Institute for Sea Exploration (IFREMER) and, above all, the exploration of the *Titanic*, which became his life's number one priority. Underlying it all was the desire to go where no one else went, to do what no one else was doing, to be the first to discover what had never been seen before.

Beyond his knowledge of the *Titanic*, he remained an aquanaut and oceanographer, whose contribution to underwater exploration and research techniques is invaluable. With him, the location and recovery of objects on the seabed – such as the black boxes from planes or helicopters after a crash – have progressed considerably. His constant vision was what had originally been developed for the *Titanic* would have many more applications in the future.

In 2022, he took advantage of the *Titanic* expedition to visit the site of a mysterious wreck nearby, spotted by sonar as early as 1998. This mass turned out to be an unknown volcanic formation at a depth of 2,900m, home to a rich and colourful ecosystem, despite there being no light at all at this depth. This underwater reef, an essential discovery to our knowledge of abyssal biodiversity, has been named the 'Nargeolet-Fanning ridge' in his honour and that of another member of the mission. For him, having discovered an exceptional natural site was even more important than having found a shipwreck.

His missions and activities were shrouded in the utmost discretion. His humility with us was such that it was the media that revealed just how well known and recognised he was throughout the world for his work. As such, it was only after his death that we learned that the Cité de la Mer in Cherbourg would never have existed without his work. He would only say that he helped the Cité de la Mer a little . . . And he would never say 'I'm the *Titanic* expert'. What's more, we didn't know that he was still going down in the submersible, we were convinced that at his age he was staying on board the escort boat. However, amongst his papers we found his latest submarine piloting certificate, dated March 2023. At the age of 77, he truly was physically and mentally fit.

Although he didn't talk much about his expeditions, he was keen to pass on his passion for the sea. Our fondest childhood memories are of regular outings with him on our *pointu*, a small, traditional Mediterranean wooden boat. Arriving at his site of choice off the coast of Sanary in Provence, he would say, 'I'll be back in half an hour', before diving in with the tanks. Once he was at the bottom of the water, he'd forget about everything else, only coming back up after an hour or more. He never asked us to accompany him on his dives, but instead taught us to pilot the boat, because he loved passing on his knowledge. One day, when we were about 4 and 5 years old, we came across an orca near Grand Rouveau Island, which was bigger than our boat.

From 1987 onwards, his life revolved around the *Titanic*. He went to sea every summer to explore the wreck, spending more and more time in the United States. Until the day he met George Tulloch, founder of the

RMS Titanic, Inc. company. This was the beginning of an incredible friendship between two men who had nothing in common. Dad was a sporty, athletic, record-breaking snorkeler, who got up every morning at six o'clock to run, swim and cycle. Tulloch, in contrast was a heavy smoker and soda-drinker. Yet their professional and friendly relationship was impressive, and when they were together they were like two brothers.

At the time, Dad spoke very little English, so dialogue between them was difficult, but they immediately became close. Thanks to George, he discovered what the United States was really like, and became fascinated by their way of life and their different vision of existence. So much so, in fact, that he moved there full-time, and then applied for, and was granted, American citizenship. He was very proud of this. In 2022, having just moved into a lakeside house, the first thing he did was plant a flagpole on the shore to hoist the American flag. While France remained his country of birth, the United States was his chosen homeland. Meeting George Tulloch, then marrying journalist Michele Marsh, were the two triggers for his move to America. The third was the proximity of the *Titanic* off the East Coast. It was their shared passion for the liner, its history and its wreck that brought Dad, George Tulloch, Michele Marsh and all his American friends together.

Even as an expatriate on the other side of the Atlantic,

he made a point of keeping in touch with his family, especially his grandchildren, of showing up for celebrations and birthdays, of being present as a grandfather, even from a distance. In fact, it was precisely to please his grandchildren and to pass on his story to them, that he finally agreed to publish this book. Until then, he hadn't seen the point, preferring to devote all his time to his explorations, which he considered a priority. When his grandchildren asked him, 'Grandpa, when are you retiring?', his answer was invariable, 'My what?'. For him, retirement didn't exist, it was inconceivable. In a way, they understood this, even if they were saddened by his death. When one of the children was told that the submersible would never come back up, he replied, 'Papy is where he always wanted to be'.

Chloé and Sidonie Nargeolet

We would like to thank Jessica Sanders, President of RMS Titanic, Inc., Bill Price from the Los Angeles Adventurers Club, James Cameron and all those who left us messages of condolence at the ceremony organized by RMS Titanic, Inc., as well as John Paschall, Renata Rojas and Rory Golden. And to the memory of George Tulloch, who played his part in introducing our father to his new life in the United States.

We would also like to thank all those who took part in the search for the *Titan* submersible, and all those who contacted us to express their support and sympathy.

Chapter 1

Saturday 25th July 1987

First dive to the Titanic

12.22. We're on target.

Three scribbled words in the *Nautile*'s logbook to express the indescribable. The bow of the *Titanic* had just appeared in our submersible's searchlights. Instantly, the cabin fell silent. Guy Sciarrone, Max Dubois and I had been exchanging information for the last hour and a half of our descent to the bottom, but no one had said a word for a good ten minutes. Before our eyes rose the incredibly spectacular view of the wreck: the foredeck, with winches and anchor chains, still in place seventy-five years later. The scene was startling. At this depth, marine life is minimal. No rippling seaweed, no carpet of sponges, no gorgonians or multicoloured corals to mask the scrap metal. The only hints of colour were the rusticles – rusty stalactites produced by microorganisms – whose orange colour contrasted with the grey veil of marine sediment and the absolute blackness of the water.

Where did the fascination with the *Titanic* come from? If the facts are anything to go by, the *Titanic* was not the outstanding liner it's often portrayed to be, nor the world's largest ship, or even the first of her line. She was one in a series of three identical transatlantic liners built for the Southampton–New York route, ordered in 1908 by the White Star Line from the Harland & Wolff shipyard in Belfast. The first to be launched, the RMS *Olympic*, went into service on 14th June 1911. Converted for troop transport during the First World War, it distinguished itself by deliberately ramming a semi-submerged German submarine, the U-103, and sinking it. After being overhauled and modernized, it resumed an uneventful career in 1920, until it was withdrawn from service in 1935. It remains known in maritime history as *Old Reliable*.

The third in the series, the *Gigantic*, was given the more modest name of *Britannic* after the tragedy. It was launched in February 1914, and was immediately transformed into a hospital ship. It sank after hitting a mine in the Aegean Sea in November 1916. Unlike the *Titanic*, the wreck is relatively accessible at a depth of 120 metres. Most importantly, its hull, lying on its starboard side, remains intact, even though the bow was almost completely detached when the mine exploded. Its ongoing exploration provides valuable information on the *Titanic*'s interior layout—in particular the boiler room and engine room, which are identical on both liners. On the *Britannic*, they remain intact, while on the *Titanic* they are unrecognizable, twisted and broken by the destruction of the first row of auxiliary boilers when the ship broke up.

The Titanic, originally conceived as *Olympic*'s twin, benefitted from its elder brother's lessons in refinement – in first class, naturally. Second and third class were more spartan, although the Titanic's third class was as comfortable as second class on many other competing liners. The interior fittings were upgraded, the upper deck was enclosed and glazed (*Olympic*'s first-class passengers had complained about the wind), and the carpets were reputedly of unprecedented thickness – details the journalists of the time raved about. And they repeated endlessly that 'the *Titanic* is unsinkable'. The company, more realistic or more cautious, actually described its liners as 'practically unsinkable', but the press ignored this nuance. The two billionaire suites with their private promenade deck attracted more press attention, as did the Café Parisien, a faithful copy of a café terrace on the Grands Boulevards of Paris. But just as the *Titanic* relegated the once headline-worthy *Olympic* to second place, so it was with the even more luxurious and grandiose *Gigantic*. Fêted with superlatives in the press, it seemed destined to dethrone its predecessor. All it took was one iceberg to turn everything upside down and render the *Titanic* unique and immortal.

Since its sinking, the *Titanic* has become the holy grail of wreck hunters, not to mention treasure hunters. Don't they say that a fortune in diamonds and ingots lies dormant in the ship's vault? In the immediate aftermath of the disaster, the craziest projects were hatched, starting with refloating the ship. They didn't know back then that it was impossible, since the hull had broken up before it hit the bottom. The first somewhat scientific attempt dates back to 1977, when geologist Robert Ballard, from the Woods Hole Oceanographic Institute in Massachusetts, tried to survey the area with detection equipment suspended from the vertical tube of a drill ship. The tube broke, the equipment was lost; it was a complete failure.

Next up was the eccentric Jack Grimm. After failing to track down Scotland's Loch Ness Monster, the Nepalese Yeti and its North American deep-forest counterpart, the sasquatch, not to mention proudly displaying what he thought was a fragment of Noah's Ark from Mount Ararat, the Texan billionaire decided to go fish for the *Titanic*'s (hypothetical) treasure in 1980. For three years, he combed a vast area around the ship's last known position, to no avail, although in 1981 he did claim to have spotted a propeller which, he believed, could only belong to the liner (he was later unable to find it). Robert Ballard, meanwhile, had also become a wreck hunter, and hadn't given up on locating the *Titanic*. A new expedition was organized in the summer of 1985, bringing together the Americans from Woods Hole and the French from Ifremer, and mobilizing the *Knorr* and *Le Suroît* vessels. The Franco-American duo Jean-Louis Michel and Robert Ballard set out to find the wreck, but to no avail. Their investigation campaign, which lasted almost a month and a half and involved the use of a new, sophisticated side-scan sonar, came to nothing.

An extra three weeks of patrols, this time using an underwater pod bristling with searchlights, sensors and cameras, the *Argo*, brought no more results. The expedition was about to return home empty-handed when, at 1am on 1st September, instead of the dreary sediment plain, a strange object loomed up on the *Knorr*'s control screens: a cylinder studded with bolts. A boiler from the *Titanic*!

That night, Jean-Louis Michel kept watch over the screens. It was only once the discovery had been confirmed that a fast-asleep Ballard was woken up. Robert later said he'd gone for a rest and was alerted by a so-called 'operations manager' who was none other than . . . the ship's cook. This didn't stop Robert Ballard from proclaiming himself the sole discoverer of the wreck and, over the following weeks, attempting to obtain full rights to it from the US Congress (a refusal he still struggles to accept today, and which has not been without consequence for the legal ups and downs of the *Titanic* case since 1985).

On the face of it, this discovery owed nothing to science and everything to chance. The *Titanic* was not where they were looking for it, i.e. around the last point estimated at 12.10am by one of the liner's officers – 41° 46′ N, 50° 14′ W. Instead, its position was 41° 44′ N, 49° 56′ W; a distance of around 14 nautical miles, or 25 kilometres. In 1985, Jean-Louis Michel's team had extended the search area to include the position of the liner *Carpathia* when it rescued the first *Titanic* survivors, but this was still very imprecise: the area to be explored covered more than 400 square kilometres. It was as if they'd been looking for a needle in a haystack . . . and by chance or by a miracle, they'd found it. However, we later learnt that a Royal Navy hydrographic vessel, HMS *Hecate*, had already located a large wreck broken into two pieces in 1980. They couldn't formally identify it, but they carefully noted the coordinates of these two metal masses so they could be used as landmarks by NATO's nuclear submarines. What's more, right from the start of the expedition an officer of the watch on the *Suroît*, Joseph Coïc, twice reported an echo that could have been a shipwreck, and during the first sonar descent, a strong magnetic anomaly was detected in the same sector. This information was then discarded as 'too good to be true'. Did the US Navy tell Robert Ballard about HMS *Hecate*'s discovery, which in theory was still classified as a 'military secret'? Did he deliberately let the 1985 expedition go astray before heading 'by chance' towards an area he knew to be promising from the outset? It's safe to assume so.

Our joy at having reached our goal on this long-awaited first dive was mixed with sadness on beholding the reality of the catastrophe.

The unfolding tragedy of the *Titanic*, or rather the version we knew of it, flashed through our minds: the iceberg, the impact which was barely perceptible to most of the passengers, the strangely calm sea, the lights flashing one last time, the liner sinking into darkness, the 1,500 victims . . . Emotions were so high on our submersible that Guy Sciarrone, our seasoned pilot, committed a slight error: the *Nautile* brushed against a railing on the wreck, and one of its safety weights came loose. Everyone snapped back into focus: we'd only been there three minutes and were already in danger of rising back up to the surface!

The start of our campaign was definitely more eventful than we expected. Two days earlier, on the *Nadir*, the expedition's support boat, we had missed the wreck. In 1987, there was no GPS. The sonar indicated we were over a geological formation known as the 'Titanic Canyon', oriented north-south. It was a dead end; we had to make a U-turn. But should we turn left, or right? Given the width of the canyon as seen on the depth sounder, we decided to turn right, trusting our gut. The Nadir headed due east again and, as luck would have it, the depth sounder soon picked up the wreck. We could now anchor and deploy the array of positioning beacons that would help the Nautile to navigate. The next day, we made a test dive to a depth of 500 metres. It was cut short by a school of whales making a terrible noise, rendering communication between the Nadir and the submersible impossible. Fortunately, the following day, the whales had moved on and the Titanic was right where we expected it to be. After that little incident, we wondered if we would manage to remain above the wreck.

Guy Sciarrone skilfully operated the ballast tank, a sphere that can be filled with water to modify the Nautile's weight. After forty minutes of hard work, the submersible, now several dozen kilos heavier, had been stabilized. We were able to go back down to the Titanic, reaching it at 1.05pm and ready to get to work. We had a long list of objectives and, given the time we had already lost, no more than four hours to achieve them. We had a map drawn up by the 1985 expedition and a few observations made by the *Alvin* submersible in 1986, but we had to verify everything. As always, the reality on the ground bore little relation to all the images and documents and there was no shortage of surprises. For example, while the *Alvin* had encountered a constant southerly current – despite the long-held belief that there were no significant currents at great depths – we, by contrast, faced a strong northerly current. And there's nothing more dangerous than a current for a submersible, which can get pressed up against the wreck and trapped by an obstacle, such as a dangling rope caught in the propeller or a heavy sheet of metal coming loose. At a depth of 3,800 metres, there's no room for error, and you're fully aware that you're on your own.

On this very first contact, we carried out a general inspection of the wreck to prepare for future dives and to identify areas of risk. We started by circling the bow, carefully ascending alongside the port-side wall, moving up to the well deck – the middle deck just aft of the forecastle. Then, we moved along the mast which was lying on the side of the gangway, to the crow's nest – the observation post from which the watchman Frederick Fleet shouted 'Iceberg, right ahead!' and rang the bell three times (in line with regulations) on 14th April 1912, at 11.40pm. From the gangway, which is covered and closed at the front but open at the sides as was customary at the time, we could see where the ship's wheel had been situated. While the wheel itself had disappeared, its base and the compass slot remained clearly visible. We then reached the captain's bedroom, which had most likely been destroyed when a funnel fell on it, giving us a glimpse of the bathroom and bathtub. There was no sign of any ropes lying around – the hemp had long since dissolved – but the mast shrouds hung loosely from the hull. Careful inspection showed these steel cables were not a danger to the submarine. The hull and superstructure looked solid enough to be approached.

We know from the testimonies of survivors that the *Titanic* broke in two just before sinking, contrary to the conclusions of the commissions of inquiry that the ship sank in one piece. But we had no idea what lay ahead as we manoeuvred towards the stern, 600 metres away. Unlike the front, which had retained its general appearance despite the heavy damage, the rear appeared to have been blown apart by a series of explosions and implosions. It was no more than a tangle of twisted scrap metal. Here and there, we spotted a few key recognisable pieces: a cylinder from a reciprocating machine, boilers and generator turbines.

Between the two parts of the ship is the first debris field. This is where we recovered the very first item from the

Titanic: at 2.37pm, the *Nautile*'s searchlights caressed the outline of a small silver plate. It would be number one in the inventory of the 1,892 objects we'd bring up between then and 9th September.

Further south, another rich debris field stretched as far as the eye could see. We soon discovered that it measured over a mile by half a mile, some 1,900 metres long by 900 metres wide. We now had a clearer idea of the terrible forces at work during the sinking. In the course of their accelerated descent – estimated to have lasted no more than five to six minutes – the two gigantic sections of the hull were partially emptied of their contents which were somehow sprinkled over almost 2 square kilometres. Each piece of hull then crashed to the rocky bottom at an estimated speed of 40-50 kilometres per hour. The violent impact caused enormous damage: the bow planted itself like an arrow in the layer of sediment and lifted itself some twenty metres to the level of the anchors; the stern sank as it pivoted, propellers forward; the central propeller was buried in the seabed, and the lateral propellers were pushed several metres up. The impact shattered the huge four-storey reciprocating steam engines, separating the first cylinders and connecting rods. Some auxiliary boilers landed several dozen metres from the stern, despite weighing 60 tonnes.

At 5.01pm, our 220-volt batteries reached their limit. It was infuriating, but we had to get back. We surfaced at

6.56pm after eight hours and thirty-seven minutes of diving; by 7.28pm we were on the deck of the *Nadir*. Mission accomplished for that day. We'd carried out this first dive before getting the official go-ahead, and the news we'd all been waiting for came while we were at the bottom. The contract between the French Research Institute for Exploitation of the Sea (Ifremer) and its American partner had been confirmed. The expedition would go ahead as planned, unlike the previous year's, which had been due to go ahead with the *Alvin* but was then cancelled at the last minute due to the lack of a sponsor. The whole team was sorely disappointed, me especially. I'd left the French Navy to carry out this mission and I regretted my decision. But this time, it was finally happening.

What I didn't yet know on the evening of that first dive was that the *Titanic* had already taken its place at the centre of my life. Thirty-four years and eight expeditions later, my whole existence revolves around it. I moved from France to the United States, where I became Director of Underwater Research for RMS Titanic, Inc. It's now a subsidiary of Experiential Media Group, the company that financed and organized the expeditions and created the exhibits of *Titanic* artefacts. I collaborated with other American companies such as Caladan Oceanic – founded by Victor Vescovo, a wealthy explorer and holder of all diving records in all five oceans – which offers dives to full ocean depths, and OceanGate Expeditions, whose comfortable five-seater observation submersible, *Titan*, takes *Titanic* aficionados on tours of the wreck. I divide my time between preparing for expeditions, diving and conferences. Somewhat amusing, for someone born in Chamonix.

My fascination with the ocean began when I was nine. My father was posted to Casablanca and I spent my time splashing around with a mask, snorkel and flippers in a pool a stone's throw from the beach. There were a whole gang training to become free divers, including a certain Lemaître, a former French artistic diving champion. I quickly spotted that they had the whole scuba diving kit. One day, I saw them getting into the water near the coast and I followed them with my mask and snorkel! Following their trails of bubbles, I saw below me a small cargo ship they'd come to explore, at a depth of 20 metres. My first wreck! I went back day after day, them at the bottom and me finning on the surface. I thought I'd been discreet, but I'd been spotted. One of the guys came up to me. I wasn't sure what to do. Did I want to try it? You bet I did! That's how I made my first dive with tanks in a Casablanca swimming pool. That led to me devouring books by Cousteau, Dumas, Tailliez and Rebikoff. The last three would later become friends.

I took up underwater fishing and taught myself to free dive like my new friends. Back in France as a teenager, after some self-training with my cousin, I did my first dive with cylinders on the wreck of a Greek-flagged ship, the *Electra*, which had sunk in northern Brittany near Roscoff. My cousin and I explored it methodically, weekend after weekend. One day, I managed to break into the lamp room and recover four copper lanterns which I triumphantly brought back. I kept two of them, and I still have them to this day. My path was already marked out: the French Navy and training as a clearance diver, followed by a career as a naval officer, mainly in mine clearance and underwater intervention.

So, when Ifremer asked me to take charge of its intervention submersibles and future expeditions to the wreck of the *Titanic* in 1986, I said yes without the slightest hesitation. I was blissfully unaware that I'd spend a year twiddling my thumbs after the expedition scheduled for that year was cancelled.

A few days before that first dive of 1987, having just arrived on site, I went up onto the deck of the *Nadir* in the middle of the night to gaze out to sea. I'm no mystic, but I was seized by a strange feeling as I gazed at the spot where the tragedy had unfolded. In a way, it all happened on the surface of the icy water at –2°C, not on the seabed.

I felt the same emotion when I discovered the wreck. It's easy to see why the *Titanic* became a legend. Far more than the loss of a prestigious liner, it was the circumstances of the tragedy that caught the imagination: the dream cruise, the stupid collision with an iceberg, the sinking on a calm sea, far removed from the spectacular shipwreck imagery of raging waves, and above all, the number of victims and their status – on its maiden voyage, the *Titanic* carried the brightest stars of the era, who at that time were billionaires, not movie stars. No doubt this sudden shift from celebration to chaos left its mark and gave rise to the wildest legends[1].

---

[1] An incredible number of books, articles and documentaries have been devoted to the *Titanic*. It's the third most written about subject, after the Bible and the Civil War.

Chapter 2

August–September 1987

Champagne and diamonds

After that first dive in August 1987, we were itching to return. The itinerary for each of the following excursions would be recorded on graph paper by the surface crew. This positioning process was still far from perfect. It relies on a regular exchange of information between the submersible at the bottom and the boat on the surface via an acoustic transmission system. But that transmission is often hampered by the sound of the *Nadir*'s or *Nautile*'s propellers, or by the calls of nearby whales, as I previously mentioned, which we discovered during the test dive when the submersible didn't communicate its position every thirty seconds, as programmed. Silences lasting several minutes or unintelligible positions were frequent occurrences.

Nevertheless, by cross-checking the results of this mapping with the observations made aboard the *Nautile*, we gained a more accurate picture of the area to be explored. The wreck rested on a bed of sediment – a mixture of sand and silt varying in colour from grey to light brown – criss-crossed by rocky outcrops. The forward section accounted for roughly half of the liner's 269 metres, so 130-135 metres, while the aft section only measured about 100 metres. Therefore, a good thirty metres of hull were missing, corresponding to the two pieces of hull found a little further on, 300 and 400 metres north-east of the wreck. The missing V-shaped section housed the first and third-class kitchens and dining rooms, some of the first-class cabins, and the first-class aft staircase. Further down lay several coal bunkers, the first part of the reciprocating steam engine room – two behemoths 20 metres long and 10 metres high, each generating 15,000 horsepower – and the row of five auxiliary boilers. The entire contents of this part of the *Titanic* were scattered across the debris field. The largest fragments had fallen near the rear, like boulders, amidst the smaller debris, including the contents of kitchens and storerooms – plates, cups, cutlery, pans, bottles . . . It was an incredible jumble of bric-a-brac – a washbasin lay adjacent to a saucepan, a toothbrush beside some steam whistles, and blocks of coal nestled next to vials of perfume. We even found a chandelier hanging from the base of one of the electric cranes.

Initial observations confirmed that there was little debris between the front and rear sections. Most of it was found to the east and especially to the south of the stern. All the light objects – bowler hats, plates and small pieces of coal, for example – had drifted in this direction with the current and could be found as far south as 800 metres away. Once they had come to rest, they did not move, remaining in a small, permanently calm zone about 30 centimetres in height, between the bottom and where the current began. This ghostly scene, frozen in time, was utterly captivating.

Thus described, the *Titanic*'s debris field sounds like a diver's paradise where you only have to reach out to grab the rarest and most astonishing artefacts. But the reality was very different. At a depth of 3,800 metres and a pressure of 380 bar – in other words, 380 kilos per square centimetre – leaving the submarine is unthinkable. Only a handful of marine organisms can withstand such depths, and it was out of the question for a human to venture outside the *Nautile*'s titanium sphere. Despite being 6 to 9 centimetres thick, dozens of litres of internal volume are lost as it compresses under pressure during the descent. The only way to collect objects is to pick them up with the pilot's remote-controlled manipulator arms and place them in baskets before bringing them to the surface. Easier said than done. This was the first time a team had worked in such conditions on a wreck at this depth. Everything had yet to be invented, tools and methods alike.

Fortunately, the *Nautile* team included some excellent and resourceful mechanics and pilots, backed up by a true mechanical genius, Ifremer engineer Pierre Valdy. On our return from the dive that evening, we aired our problems and expectations. The next day, the *Nautile* had been fitted with a new tool, designed and built by Pierre and the mechanics overnight. Ahead of the *Titanic* expedition, some disappointing attempts were made to pick up plates and fragile objects with the manipulator arm's hydraulic clamp. The clamping action is very difficult to control: nine times out of ten, the plate breaks . . . Fortunately, they were *Nadir* plates sacrificed by our cook and not the *Titanic*'s precious historic plates. But a solution was found: equip the arm with a suction pad connected to a vacuum pump. Thanks to this device and a whole arsenal of improvised tools (ladles, forks, pitchforks, hooks, shovels, etc.) laid out on a rack in front of the *Nautile*'s basket, we managed to gently remove the china, carafes and many other items, and bring them to the surface intact.

Picking up the artefacts was one thing; bringing them to the surface without loss or damage was quite another. Once again, creativity was the order of the day. At the very front of the *Nautile* is a small, watertight, isothermal 'scientific' basket, originally designed to bring up live marine animals. This water-filled basket – water being the best shock-absorber there is, once the basket is closed – proved ideal for conserving small, fragile objects. This was especially true when hoisting the *Nautile* aboard the *Nadir* in rough seas, which are not uncommon in the North Atlantic.

For big catches, it's a different story. The team, with unfailing ingenuity and adaptability, came up with the 'basket elevator' system. Some had already been designed for scientific missions, or to recover aircraft debris from previous missions. Depending on the type of object to be lifted, these are either open baskets made up of an aluminium tube structure lined with nets, or lidded baskets made up of two waste containers installed back-to-back in a frame also made of aluminium tubing. A foot pedal opens the lids, which are closed again before the baskets are returned to the surface.

The elevator system was as simple as it was ingenious. The baskets, fitted with floats, are launched and sink to the bottom with a descending ballast. Weighing and preparing weights, sinkers, etc., is an art. Christian Le Guern, the *Nautile*'s chief engineer, was a master of that art. The pieces of chain and the shot bags had to be juggled, and then arranged to allow the *Nautile* to use the knife attached to the arm clamp to cut the links gradually, until the whole thing was correctly weighed and could be transported to the right place. When a basket is dropped from the surface, you can't be sure where it will land 3,800 metres below. The current often changes from day to day, even during the course of a dive, and this applies from seabed to surface. Once the *Nautile* had filled the basket with artefacts, the ballast was released by an acoustic signal and, by virtue of Archimedes' buoyancy, the floats raised the precious cargo. Obviously, we had to avoid overloading.

On the surface, divers took charge of the baskets. Open baskets were covered with a net that had been rolled to one side to prevent items from slipping out when hoisted aboard, while lidded baskets were hoisted aboard directly. All were equipped with homing beacons – a pinger identical to those fitted to aircraft black boxes – emitting a 38-kilohertz acoustic signal, and a flashing light. Given the short range of the *Nautile*'s searchlights and the size of the area to be prospected, this equipment was essential – no one wants to waste time searching for baskets in the middle of a debris field[2].

---

[2] During the *Titanic* missions we learnt just how difficult it can be to detect pingers in the midst of debris, even when placed a few metres above the seabed, as their range was far shorter than the manufacturers claimed. In theory it's easy; in practice it's a different story.

We were very proud of our little suction cup system. It meant that none of the 5,500 or so items brought up by our successive expeditions were damaged or lost. On 11th August, it scored us a great find. That day, following a false move on the surface by the expedition's support vessel, the *Abeille Supporter*, a basket, already stripped of its floats but still fitted with its transponder, fell back into the sea. Thanks to its beacon, we located it some 600 metres north-east of the wreck's stern. The next day, the *Nautile* searched for and found the basket. After re-equipping it with floats to return it to the surface, the submarine headed for the stern of the *Titanic* through a previously unexplored area. This is how, by complete chance, we discovered the long-lost first half of the double hull, separated from the two main parts of the wreck. At first glance, the piece seemed insignificant. It lay flat on the bottom, with the outside part facing up. The double hull, anti-roll keels and central keel were visible. But it would be very useful to naval architects in reconstructing how the hull broke up, especially when we discovered the second missing piece in 1993. After discovering the double hull, we launched the *Nautile* far enough away from the wreck to explore the surroundings as thoroughly as possible, so as not to leave any interesting areas unexplored.

* * *

Working underwater is no picnic, since visibility varies greatly from one dive to the next. Even on days when the water was fairly clear, we were in the Gulf Stream zone which descends to 400 metres in this area and is full of plankton and particles. When these particles leave the stream, they descend in flakes that reflect the light from the searchlights. On days when they fell on the wreck, it was like sailing through a snowstorm. The same difficulties arose with the vagaries of the underwater current. When it reversed, it 'washed off' the sediment deposited on the wreck, creating a cloud effect that impaired visibility.

Our submersible could raise clouds of sediment itself if one or other of its vertical motors stirred the water downwards. Normally, the current means you regain visibility quickly, but sometimes there's no current at all. You either have to move to another area or remain stationary for twenty to thirty minutes, until the water is clear again. The *Nautile* weighs 18 tonnes, and its weight on the bottom is set to be neutral. Care must be taken not to weigh the submarine down. It's best if it's sufficiently light to avoid using the vertical engines to ascend, as this would risk stirring up the sediment and reducing visibility. And, of course, there's no brake pedal. The pilot's art lies in approaching an object slowly, arriving at just the right distance to grab it, but not go past it. Otherwise, you'd have to use the propeller to reverse, with the risk of blurring visibility. We quickly learnt that when we spotted an interesting object in the jumble of the debris field we had to get it right away, otherwise we'd either never see it again or waste an inordinate amount of time searching.

With each dive, the inventory looked more and more like a random jumble: the captain's megaphone, the cherubic torchbearer (minus the torch) on the aft first-class staircase, a 650-page dictionary that brought the curator on board the *Nadir* to despair, each page requiring a whole day of restoration . . . On several occasions, we found strange bottles, neatly arranged at the bottom. The wooden box containing them had gone, but they still contained a milky substance. There was no way it could be milk, which would never have kept for so long. We decided to bring up a bottle and, after just a short time at the surface, it transformed. The mysterious contents become liquid again, regaining their original colour and revealing their true nature: cooking oil, frozen by the low temperatures. Strangely enough, no one tried tasting it . . .

We learnt the hard way that some seemingly intact items were no more than ghosts. After bringing up several pairs of binoculars on previous dives, one day we discovered another pair, lying prominently on the sediment. The second the manipulator arm touched it, poof! A small cloud, and then nothing. Once the shock had passed, we soon found the culprit: the acidity of the sediment, which shows up between 2,000 and 2,500 metres and increases with depth. Here, the pH is 4. If you touch the sediment (which sometimes rises to the surface in the submarine chassis, or coats certain items), your fingers will immediately prickle and the tips will turn smooth for one or two hours. The combination of acidity and sea salt gives off chlorine, which slowly attacked the copper in the binocular tubes, thinning them until they crumbled into powder as soon as they were moved. By contrast, if a copper object comes into contact with a piece of iron, an electrolytic couple forms, with the iron acting as a sacrificial anode and protecting the copper. If iron isn't present, the slightest touch will cause it to evaporate as if by magic.

At the top of our list of items to be recovered was a safe seen in photos taken by the American submarine *Alvin* the previous year. Given what was at stake, improvisation was no longer on the agenda. Before leaving Toulon, the inimitable Pierre Valdy had built a clamp capable of handling the safe, whose maximum weight was estimated at 1 tonne. Tests carried out on land with an even larger and heavier concrete block were conclusive. Full of foresight and never short of ideas, Pierre had also developed a tripod lifting system to enable the safe to be placed in a net before being brought up. While it was fire-resistant, thanks to a thick layer of firebrick, the safe's thin metal walls were badly corroded. The self-tightening clamp might damage it . . .

In the end, the 'monster' turned out to be much lighter than expected. We wrapped a rope around it, lasso-style, and the *Nautile* lifted it up. The operation worked the first time. It was placed in a basket and brought to the surface with the utmost care. There was no way we were going to lose it. The contract between the American clients and Ifremer included a $200,000 bonus for bringing up a safe. At first glance, obvious marks on the door revealed that, despite Robert Ballard's solemn undertaking not to touch anything on the wreck, the crew of the *Alvin* had tried and failed to force it open. At least they hadn't damaged the bronze handle. Later, when we opened it in front of witnesses on a TV show in Paris, we understood why the safe was so light. It was empty . . . except for a leather bag containing coins. We concluded that it must have been a third-class safe.

A bag that didn't look like much at first glance had a surprise in store. Several of these leather 'Gladstone bags', long used by doctors on a daily basis, lay at the bottom of the sea. One, in particular, caught our eye from the start, as both its handles had been sliced clean through as if with a knife. It had been decided at the start to stick to items belonging to the ship – we were to try and refrain from touching personal belongings to avoid fuelling controversy around respect for the victims. But identifying these items amidst the sediment wasn't that simple, not to mention the fact that curiosity always wins out in the end. The day the submersible resurfaced with this bag I was on deck to collect and do an inventory of the dive's yield. I opened the bag, and started by taking out a man's white shirt rolled into a ball, a white glass bottle, a small copper plate with the name of a cash-in-transit company, and then, underneath, a pile of small boxes, which turned out to be jewellery cases. At the very bottom were wads of banknotes and gold coins. Was it the contents of a safe hastily transferred by a purser? Why the sliced-through handles? It was a mystery. All we knew was that the jewellery items bore different initials, proving they did not belong to a single person.

We were intrigued by a double box. Each of its two parts contained a rather chunky, solid gold necklace, with a large nugget of pure gold dangling from it. The two necklaces were absolutely identical, as were the two nuggets. This was no accident. In fact, the lighter one of the two turned out to be a faithful copy of the other . . . Number 103 in the 1987 inventory, this bag contained the vast majority of the jewellery items found to date on the wreck and remains the *Titanic*'s only known treasure. Among the finest pieces were Victorian and Art Nouveau rings and pendants – in particular a very fine ring set with a splendid sapphire surrounded by diamonds. The bag also contained a rare treasure very much of its time. After restoration, some of the notes showed the words *silver dollar* – i.e. redeemable in silver coins of the same value – and bore the name of a bank, testimony to an era when the American government left it up to the states and private banks to issue their own banknotes. The diversity of the specimens makes this a rather unique collection.

While the majority of objects were retrieved without difficulty, the suction cup-basket system working satis-factorily, we faced a major challenge in the shape of a miraculously intact stained-glass window, presumably from the library or the first-class smoking room. Measuring around 150 centimetres high by 40 centimetres wide, it had bent to match the undulations of the seabed on which it had rested for over seventy years. There was no way of knowing how much the lead had corroded, but it was certainly extremely fragile. We couldn't risk lifting it. The solution was to make a large shovel of similar dimensions, which was gently slid into the sediment beneath the stained glass.

Once on the shovel, it was covered with a plywood lid lined with shock-absorbing foam on the inside and clamped in place with concrete reinforcing bars. Then all we had to do was place the 'sandwich' in a specially enlarged basket, and head for the surface. One dive was all it took to carry out this near-surgical operation. On 21st August 1987, the *Nautile* left at 8.38am and arrived at the stained-glass window at 11.40am. At 12.48pm, the stained-glass window was placed in the basket, and released at 2.59pm. The surface team retrieved and unwrapped it with infinite care. All the parts were still in place and the lead hadn't been too badly damaged by the acidity of the sediment. When we unpacked it on the deck of the *Nadir*, the stained glass was still lying on the shovel with the sand and silt. It bears witness to both the refinement of the *Titanic*'s interior design and the company's desire to make first-class passengers forget they were at sea. Everything was designed and organized to give the impression of staying in a grand hotel on land throughout the crossing.

On board the *Nadir*, basket-raising was a popular attraction. Everyone came to see the day's catch and comment on the finds. Once, we brought up some simple, eared egg dishes in white porcelain. On the back was the manufacturer's monogram and the word 'Paris'. The ship's cook couldn't believe it: on the transatlantic liner *France* he'd used exactly the same ones, from the same manufacturer . . .

Once the items had been inventoried, measured and photographed, preservation measures were applied to prevent their rapid deterioration in damp, open air. This was particularly true of metallic objects, which oxidize rapidly. Often, iron cannonballs that dried out shortly after leaving the sea ended up shattering into pieces.

On board, objects were stored in bins of fresh water, sometimes wedged in with blocks of the compact foam used by florists. Back at port, these bins were transported to laboratories specializing in conservation. The diversity of objects recovered from the *Titanic* has led to great strides being made in this field. The EDF laboratory, which used electrolysis for metal objects, perfected the electrophoresis process. It makes non-conductive objects conductive and then treats them with electrolysis. But conservation isn't permanent, especially when it comes to ferrous metals. It's very difficult to get to the heart of the object, even with a chemical treatment, and corrosion can resume, requiring some objects to be treated repeatedly and regularly. A very effective treatment had just come out which involved immersing the artefacts in a neutralizing solution, then placing them in a small hyperbaric chamber[3] and increasing the temperature and pressure until they behave like both a gas and a liquid. The gas phase penetrates deep into the molecules, while the liquid phase dissolves

---

[3]   A sealed enclosure in which a pressure greater than atmospheric pressure can be created. Hyperbaric chambers increase tissue oxygenation, particularly in the event of diving accidents or carbon monoxide poisoning.

salts and chlorides. This treatment is permanent and very fast, but still very onerous.

Non-ferrous metals are less problematic: the more noble they are, the more resistant they are. Gold and platinum are unaffected, silver and stainless steel are highly resistant, lead a little less so, while copper and bronze are only sensitive to acidity. The bronze bell from the Titanic, for example, is in fairly good condition, except for the part in contact with the highly acidic sediment, which has been eaten away. This shows beyond any doubt that we recovered it from the debris field and not from the crow's nest on the main mast, as rumour had it.

Fabrics, paper and leather barely deteriorate at all once treated. Thanks to chemical impregnation during tanning, leather stands up well, and leather containers – such as bags, saddlebags, suitcases and wallets – have protected the paper documents they contain, including the ink used on letters that are still legible. Porcelain is easy to clean, and any rust stains created by contact with metal parts disappear when chemicals are applied. Glass remains transparent, and the crystal decanters, once cleaned and rinsed, sparkle like new. To best preserve them, all these items go on public display in showcases where temperature and especially humidity – the great enemy of treated objects – are controlled and regulated. Thanks to these techniques, items from the *Titanic* are gradually being restored to their former glory and given a new lease of life. They were the silent witnesses to the comings and goings of the hundreds of passengers aboard the ship, whose lives hung in the balance on that fateful day in April 1912. The wreck is a time capsule, and the artefacts are its historical memory.

On 9th September the 1987 expedition came to an end. We far exceeded our objectives: twelve dives were scheduled in the contract, thirty-two were carried out and 1,892 items were brought up. The American team were delighted and would be back to work with us again. Strong bonds of friendship had formed between the teams.

In addition to the figures, we learnt a great deal and made great progress in our understanding of the wreck and how to salvage its remains. We also got our fill of memories and sensations. Like the day I took the first bottle of champagne out of the *Nautile*'s basket. It looked full, and the cork, while badly damaged, was still in place. Suddenly, decompression released and the gas dissolved in the bottle, making a pleasant smell of champagne spread throughout the boat. That evening, we breathed in the ultimate *Titanic* fragrance.

Many of the questions I get asked today concern life aboard a submersible: how do you cope with spending up to twelve hours in 0°C water? Can you breathe comfortably? Is special training required?

Regarding the cold, it depends on the submersible! The shell is usually made of steel or titanium, which are roughly the same in terms of thermal conductivity. As a result, the inside surface is always very cold. But in the *Nautile*, the many electronic devices give off enough heat to warm the atmosphere and stabilize the temperature at around 15°C. What's more, the crew is insulated from the sphere itself (i.e. the crew cabin, which offers a volume of 4^{m3} of air for three people) by the interior frame which the equipment, seats and bunks are attached to, and air, as we know, is the best insulator. That said, when you sit still for hours at that temperature, you eventually get cold. We don't wear special outfits equivalent to wetsuits; we prefer thermal underwear, thick fleeces and neoprene shoes.

In 2021, we dived aboard the *Titan*, which has a different design. A carbon-fibre structure connects two titanium hemispheres, one containing the equipment, the other housing the main porthole. When in direct contact with the titanium, it's freezing cold, while the carbon-fibre section, where the pilot and passengers sit, stays at a pleasant temperature because its walls have a lining which holds the electrical cables.

What we breathe in the submersible is just normal air at normal atmospheric pressure, nothing like the gas mixtures in the cylinders used by scuba divers to combat disorders such as nitrogen narcosis[4][5]. This air is recycled by removing carbon dioxide, which is absorbed by agents that fix it. These agents are either sodium or lithium. Soda lime, also used in large military submarines, comes in granulated form and has the advantage of being inexpensive. Lithium, which is more expensive but has a high fixing capacity, is very easy to use. Just sprinkle it on the ground or spread blankets containing it, and there's no need to pass gas through it to fix it. In general, lithium is used as a last resort for safety, in case the submersible gets stuck at the bottom. While the carbon dioxide is being removed, oxygen is re-injected to maintain levels of around 21 per cent. Any more would increase the fire risk; any less could trigger hypoxia or reduced oxygen levels in the blood. This leads to unconsciousness (without symptoms) within seconds and occurs below a level of 17 to 18 per cent. To enhance safety, the on-board instruments display oxygen and carbon dioxide levels continuously and in real time.

And then there's the thorny issue of toilets. To go ten or twelve hours without relieving yourself would

---

[4] Inner lining designed to insulate against direct contact with the hull.

[5] At depths in excess of -40 metres, the pressure of the nitrogen contained in the compressed air becomes toxic, affecting divers' nervous systems. It's known as the rapture of the deep.

be superhuman. In the *Nautile*, the facilities were frankly rudimentary, with everyone discreetly making do with bottles and other containers, and the crew resorting to an abundance of nappies. In the *Titan*, designed to accommodate passengers who have paid (very dearly) for their trip, there is a real toilet concealed by a curtain, which you can sit on. It's the only submersible I know of to offer that kind of luxury. When I had the opportunity to dive with the Japanese, I discovered that they had adopted the system used by fighter pilots: a small flexible pouch containing granules that instantly crystallize urine through a chemical reaction. It's a lot more elegant than our bottles . . . But whatever the solution, it's best to avoid drinking in the twelve hours preceding the dive. In truth, this is the only real limitation. Otherwise, deep-sea diving in a submersible is the only extreme activity accessible to anyone in good health, with no training and no age limit. But you need to be able to stay in a confined space for several hours without becoming stressed. On this score, there are parallels between a submersible and a space capsule. When he was invited to dive to the *Titanic* aboard the *Nautile* on 28th August 1996, Buzz Aldrin, the man who walked on the Moon, declared, 'It's a timeless experience; an unforgettable moment. And it's a lot like piloting a spaceship. In short, it's the most exciting thing I've done since my walk on the Moon.'

Chapter 3

1989

Ghosts, fantasies and forgers

In the spring of 1989, a journalist called me. He'd just read the proofs of a book containing passages about the *Titanic* that left him perplexed, and he wanted my opinion. When I read the proofs, I nearly fell off my chair. The author began by asserting that bodies had been found inside the wreckage. He then described a fantastical scene in a lounge, where passengers were still standing up in their evening clothes, the men in tuxedos, the women in long dresses . . . A veritable dance of death. All this information was said to have come from a senior Ifremer official whose job compelled him to remain silent, and who handed the author the mission of revealing it. All in all, a lot of nonsense.

I happen to know the author, a military doctor and adventurer – not necessarily in the best sense of the word.

He was involved in the sinking of the *Rainbow Warrior* in Auckland in July 1985 by the French secret services. I went on a fact-finding mission to Ifremer and ended up receiving a call from the institute's managing director, a man with a great sense of humour whom I'd met several times. He told me the end of the story: annoyed by the doctor's boasts at a Paris party about his alleged exploits around the Rainbow Warrior, he came up with this nonsense story about the *Titanic*'s ghost passengers, simply for the pleasure of shutting him up. This is how the 'information' found its way into the book. So, I called the author, with whom I'd already been in contact, to tell him what I thought. I asked him how a doctor like himself could imagine for a second that bodies could be preserved in a good condition for over seventy years in water whose temperature is well below freezing point (0°C on average). In the end, happily, the book was published without the offending passage.

This colourful anecdote is just one of the many avatars of the myth of the *Titanic-tomb* in which the 1,500 victims of the sinking are said to rest. This ancient myth was given a new lease of life with the discovery of the wreck. It took root in Internet forums, and in the statements of certain officials. To understand why, a brief look back is in order.

* * *

Episode 1: It all began with the 1985 expedition. The contract between the Woods Hole Oceanographic Institute and Ifremer gave the latter exclusive rights to images taken of the wreck, including selling them 'to the rest of the world', i.e. outside the US. For Ifremer, the sale of the images would be used to finance the French partici-pation. As the Woods Hole operation was financed by public funds, the images were, by law, copyright free. The *Titanic* expedition was part of a mission whose equipment was financed by the US Navy but developed by Robert Ballard's laboratory. The supposedly confidential mission was to test and prove that the equipment was capable of finding and photographing the wrecks of two US Navy nuclear-powered attack submarines. The *Thresher* (SSN-593) and the *Scorpion* (SSN-589) crashed in 1963 and 1968 respectively, and the US Navy knew their exact position. Once the two submarines had been found, the Secretary for the Navy authorized the expedition to the *Titanic* in order to continue testing the equipment. If successful, this would have been a clear demonstration of the US Navy's ability to find and photograph anything in deep water, a message directly addressed to the Soviets.

As soon as they returned to port, the precious images, recorded on physical media – the Internet didn't exist at the time – were flown to France by the first plane.

Robert Ballard, not one for fair play, quietly released the same images to the American press before they arrived in France. Due to the time difference, they were published before the French images, which were then worthless. Ifremer were furious but took it on the chin. Woods Hole felt similarly, and its director later formally banned Robert Ballard from recovering any items on the next expedition with Ifremer, which was already being organized.

Episode 2: With an aura of prestige as discoverer and photographer of the wreck, on 29th October 1985, Robert Ballard appeared before the United States House Committee on Merchant Marine and Fisheries. The legislation drafted to protect the *Titanic* site was later signed by President Ronald Reagan. Ballard presented lawmakers with a summary of the discovery and demanded the right to bring up objects and to film inside the wreck, in order to preserve a lasting, tangible record of the *Titanic* for future generations. He clashed with Jon Hollis, spokesman for the Titanic Historical Society, who demanded that the wreck site, the final resting place of the 1,500 victims of the sinking, be protected by law from treasure hunters. He said their presence amounted to sacrilege ('[. . .] *not to create a sacrilege by allowing purveyors of profit to desecrate this gravesite'*)[6]. Hollis' case was supported by numerous testimonials from *Titanic* survivors, maritime museum curators, marine antiquaries, etc. Ballard, on the other hand, was only supported by the colourful Jack Grimm, whose appearance before the commission was highly unconvincing. He insisted, under oath, that he had spotted the Titanic as early as 1981 and added that Woods Hole and Ifremer copied the equipment he had developed and financed for the search. In the end, Hollis won, and the house committee's text classified the Titanic as an international maritime memorial, to be left undisturbed until rules for site surveys and possible salvage had been established. Robert Ballard had to bow out. However, he took the opportunity to do a U-turn before his audience and the media. If he, the discoverer of the Titanic, couldn't, as he had envisaged, bring artefacts up so that 'a part' (sic) could be presented to the public (I suppose he was planning to sell the rest) then nothing at all should be salvaged, and Ballard added that anyone doing so was a grave robber. He had orchestrated a controversy about bringing up items prior to the 1986 expedition with the

---

[6] The full text of the speeches and facsimiles of the documents were published in a 116-page booklet entitled *Titanic Maritime Memorial Act*, available online at https://www.gc.noaa.gov/documents/hr3272-house_merchant_hearing.pdf.

Alvin, his mini-ROV[7][8], the Jason Junior, and in partnership with Ifremer, the Nautile and its mini-ROV, the Robin. The dives were planned with two submarines, the Nautile, which could recover objects, and the Alvin, which could not.

Episode 3: 1986. So, Robert Ballard led an expedition to the wreck. A few weeks before the start of the expedition, Ifremer pulled out as it had been unable to source funding outside the budget allocated by the French government. This budget was earmarked for scientific operations, but, as an Epic (an industrial and commercial outfit), the Institute had to supplement it with external income from operations, patents, and scientific collaboration with other countries . . .

We were on the *Nautile* in the middle of an expedition with geologists off the Spanish and Portuguese coasts when the bad news hit. All our dreams of diving on the *Titanic* evaporated. And all the efforts of Ifremer management in Paris and the technical teams at the Toulon centre, including Jean-Louis Michel, who had worked so hard to put this project together, were dashed.

---

[7] Jon Hollis changed his position in the early 1990s, declaring, *'I was wrong [. . .] I now realize that the objects should be preserved for the public in a museum where they can be viewed.'*

[8] *Remotely operated vehicle.*

Officially, Robert Ballard was polishing his image as a man who respected the decision of the US Congress and the memory of the dead by placing two plaques at the site. One, at the stern, bears the inscription: IN MEMORY OF THOSE SOULS WHO PERISHED WITH THE 'TITANIC' APRIL 14/15 1912. DEDICATED TO WILLIAM H. TANTUM IV[9], WHOSE DREAM OF FINDING THE 'TITANIC' HAS BEEN REALIZED BY DR. ROBERT D. BALLARD. THE OFFICERS AND MEMBERS OF THE TITANIC HISTORICAL SOCIETY INC. 1986.[10]

The other, on the bow, proclaims: 'THE EXPLORERS' CLUB'. IN RECOGNITION OF THE SCIENTIFIC EFFORTS OF THE AMERICAN AND FRENCH EXPLORERS WHO FOUND THE RMS TITANIC: BE IT RESOLVED THAT ANY WHO MAY COME HEREAFTER LEAVE UNDISTURBED THIS SHIP AND HER CONTENTS AS A MEMORIAL TO DEEP-SEA EXPLORATION. BOARD OF DIRECTORS, 4TH JULY 1986.[11]

---

[9] William Harris Tantum IV (1930-1980), co-founder of the Titanic Historical Society, died before he could realize his lifelong dream of diving on the wreck.

[10] 'In memory of the souls who perished with the *Titanic* on 14th and 15th April 1912. Dedicated to William H. Tantum IV, whose dream of discovering the Titanic was realized by Dr Robert Ballard. The Office and Members of the Titanic Historical Society, 1986.'

[11] 'The Explorers' Club'. In recognition of the scientific efforts of the American and French explorers who discovered the RMS *Titanic*: Be it resolved that whoever comes here at a later date leave this vessel and all it contains in peace, as a memorial to deep-sea exploration. The Board of Directors, July 4th, 1986.'

A clever PR stunt, with images widely published in the media. Unofficially, the 'leave in peace' rule seemed to be applied rather elastically – just think of the marks left by Robert Ballard and the *Alvin* on the famous safe.

Episode 4: Summer 1987. A new Franco–American expedition, this time bringing together Ifremer, Taurus International, Westgate Entertainment and the recently formed Titanic Ventures (future RMS Titanic, Inc.), founded and headed up by George Tulloch, was about to descend on the wreck to recover artefacts from the debris field. When Ballard found out, he was furious. He did everything in his power to torpedo Ifremer. From a legal point of view, it was a dead end. The wreck lay in international waters and Ifremer is a body governed by French law. So he used the only means at his disposal: mobilizing public opinion.

Reneging on his official declarations of 1985, he did an about-face, adopting Hollis's thesis of the *Titanic* as the final resting place of the 1,500 victims, and exploration as a violation of a burial ground. His goal was to arouse the indignation of survivors of the disaster, in particular Eva Hart, a shipwreck survivor whose father disappeared in the sinking when she was just seven years old.

Eva couldn't find words harsh enough for 'the treasure hunters, vultures, pirates, grave robbers'. Her declaration – 'This immense tragedy leads me to consider the *Titanic* as a grave. The ship is a memorial. Leave it where it is.' – was widely circulated by opponents of its recovery. However, Ballard's campaign failed, the exploration went ahead as planned, and Ifremer brought all the artefacts back to France to be processed by the EDF foundation in Saint-Denis, north of Paris.

That was the end of the story, but the controversy didn't go away. On 21st January 2020, Nusrat Ghani, then British Minister for the Oceans, welcomed the coming into force of a *Titanic* protection treaty signed in 2003 between the United States and the United Kingdom: 'This momentous agreement between the USA and the UK to preserve the wreck means that it will now be treated with the sensitivity and respect due to the final resting place of 1,500 people.' A month later, on 20th February, Ciarán McCarthy, an Irish maritime law expert opposed to the recovery of the Marconi telegraph, declared: 'It is a matter of intense concern that the hull of the wreck, which is essentially the grave of more than 1,500 passengers, should be penetrated to facilitate the removal of relics.'

The idea of the *Titanic* sinking in one piece and dragging into the abyss all those who could not find a place on the lifeboats, a notion rubber-stamped by the conclusions of the commissions of inquiry and popularized by the illustrations in the press of the time, does not, however, stand up to scrutiny of the archives, nor to exploration of the wreck. The reality was a broken hull that largely emptied its contents as it sank and belies this assumption. Only warships are likely to become mass graves in the event of shipwreck when all openings are hermetically sealed at the 'zero stage' of battle stations. This was, for example, the case with the Japanese battleship *Yamato*, which sank in 1945 with more than 3,000 men trapped inside[12].

On a cruise liner, it's the other way around. In the event of an alarm, everyone is instructed to go on deck and put on a life jacket to have a chance of surviving, even when the lifeboats are full. We know from eyewitness accounts that armbands were widely distributed on the *Titanic*, including to third-class passengers. We also know, again thanks to the survivors, that the mechanics and drivers in danger of perishing behind the watertight doors escaped because the doors were quickly reopened. But very few of them, and none of the engineers, survived. It's possible that some passengers and crew remained on the ship when it sank, but these unfortunates represent only a tiny minority. This is not to minimize the *Titanic* tragedy,

---

[12] I had the opportunity of working with a Franco-Japanese team including a few survivors to bring up artefacts from the wreck which are now on display in the maritime museum in Kure, where the battleship was built.

of course, but to point out that sensitivities and reactions to it can differ, while still being valid.

Those shipwrecked on the *Titanic* didn't drown, they froze to death. In water as cold as –2°C, as was the case that night, life expectancy is very limited. Hypothermia inexorably gains ground, and the fatal outcome occurs between five and twenty minutes after falling into the sea, depending on the individual's hardiness and the clothes he or she is wearing. When the *Carpathia* arrived on the scene at 4am, a hundred minutes had elapsed since the *Titanic* had sunk, and there was no chance of finding any survivors. Witnesses described the sea as a field of corpses floating on the surface in their life jackets.

In the days that followed, the White Star Line chartered several ships to recover the bodies. On 17th April 1912, the cable ship *Mackay-Benett* set sail from Halifax, soon joined by the *Minia*, *Montmagny* and *Algerine*. Among them, they collected around 330 bodies before calling off the search on 8th June. In this often-turbulent area, many victims drifted fast and far, pushed by winds and currents, and could not be located. On one of our expeditions to the area, we lost a large, bright-orange tank containing 20 cubic metres of diesel – a far cry from a human body floating on the surface. The two-boat search began within half an hour and lasted twenty-four hours, to no avail. Drifting aimlessly, this 'big bag' was never found.

But the *Titanic-tomb* fantasy is still very much alive. To this day, one of the most frequently asked questions is whether we found any bodies in the wreck. The answer is no. We saw neither bodies nor bones. Although it's not impossible that a few victims may have been trapped inside the boat, the acidity of the water has long since dissolved all organic matter. Some people, however, will always see what they want to see, claiming that the leather shoes (preserved by tanning) found scattered on the debris field are the final traces of missing bodies. Following the 2010 photographic and mapping campaign, an archaeologist released a photo of clothing, claiming that it belonged to a skeleton. On further investigation, it turned out he'd cropped the image, which simply showed an open suitcase with clothes and objects spilling out.

Since the deaths of the last survivors of the shipwreck, the tide of public opinion has turned. Today, a majority approves the recovery of the *Titanic*'s artefacts, considering them to be the ship's tangible historical memory. But macabre stories continue to flourish on the Internet, such as the one where 'the only human debris recovered was a phalange bone brought to the surface in a metal tureen dish, protected by the oxidation of the victim's wedding ring'. Two dishes were lifted before my eyes, but neither contained anything but sediment – besides, what would a finger be doing in a tureen dish? And let's not forget that no metal can prevent acid destroying organic matter. The formation of a protective galvanic couple only occurs between two metals of different natures. It's an unbreakable law of physics, but man's formidable imagination can refuse to stop at the bounds of physics . . .

Imagination also knows no bounds when it comes to selling artefacts supposedly from the wreck of the *Titanic* to collectors with more money than sense. And doing so, despite their sale being forbidden by the decision of a US federal judge in the Norfolk court in Virginia. The alternative is simple: either these artefacts were stolen, or they are fakes. Stolen artefacts, usually in the form of debris, may have been removed from the support vessels by employees of expedition partner companies, or misappropriated by dishonest collaborators. Strangely enough, the same names crop up regularly in transactions involving dubious artefacts and documents on confidential websites. Adverts usually combine true and false information to gain potential buyers' trust and lull them into a false sense of security. Often, they are all too willing to believe what they're told. One item for sale was a fragment of the *Titanic*'s hull from the 'Big Piece' – an authentic 20-tonne piece of hull brought to the surface by our teams in 1998. This fragment of rusted metal, 'an exceptional relic . . . guaranteed 100% authentic' (sic), is said to have been

recovered from an accident involving the truck transporting the Big Piece in 2001. It was displayed in a plain
white cardboard box (given the asking price – 1,800 euros
– one might expect a slightly more sophisticated case) along
with an impressive certificate of authenticity signed by
the former Director of Operations of RMS Titanic, Inc. On
further investigation, it turned out the signatory did in fact
take part in two expeditions on the *Titanic*, as an employee
of a service provider, in 1993 and 1994. He was responsible
for the security of the artefacts (you see the result). He then
joined RMS Titanic, Inc on the 2000 expedition[13], but was
quickly dismissed. He was never an operations manager
and never had the authority to sign anything, least of all
for an item from an expedition in which he had no part.

When contacted, the three people in charge of exhibitions at RMS Titanic at the time said they had never heard
of any accident while transporting the Big Piece to the
exhibition site in 2001, nor in any other year. According
to the ad, the seller worked with a company specializing
in ocean-liner memorabilia, selling a mixture of authentic
items and others of dubious origin. He was said to have
been Director of Foreign Exhibitions for RMS Titanic, Inc,
a position that never existed. Another vendor claimed that
George Tulloch gave him items to thank him for his

---

[13] See chapter 8.

outstanding service to the company. He even wrote a letter attesting to this and tried to forge George's signature at the bottom. Knowing this character and what George thought of him, it's enough to make me hit the roof.

In the same vein, I was recently asked to authenticate a photograph of keys, apparently from one of our expeditions. They were described as belonging to the binocular case from the *Titanic*'s crow's nest, but we never recovered anything like them. As far as I'm concerned, these two keys with their pretty bone label are fakes – the bone wouldn't have withstood the conditions at the wreck site. They were nonetheless sold as authentic.

Other small items left in the pockets of survivors, or found on the bodies of victims, have fetched record prices – more than 55,000 euros for a sailor's locker key. Even documents with no connection to the sinking have become eye-wateringly expensive. A profile drawing of the *Titanic*'s hull fetched almost 270,000 euros at an auction in 2010. All in all, there's plenty here to motivate many a crook . . . The only authentic metal from the *Titanic* and officially put up for sale is from steel samples taken in 1993 by Russian submersibles during the filming of the 1995 Imax documentary *Titanica*. The metal was used for laboratory resistance tests before being bought by a Swiss watchmaker to make the dials for a limited series of luxury watches.

* * *

As a general rule, the utmost scepticism is called for when it comes to objects allegedly found in prestigious shipwrecks. Any crockery marked 'Titanic' is a crude forgery. For reasons of economy, the porcelain or earthenware on board, common to all the company's liners, was simply stamped with the inscription 'White Star Line', surmounted by the red flag with the white star. Certificates of authenticity are only valid if signed by those with the authority to do so.

I've met several collectors capable of spending a fortune on a small piece of paper, such as a letter written by a passenger on the *Titanic*. I remember one in particular, a multi-collector of artefacts, each more extraordinary than the last. Antique cars, giant jade statues, Native American artefacts . . . His home was a veritable museum, and the room devoted to his collection of objects from the *Titanic* and the White Star Line was impressive. He was far from alone in having the means to finance his passion, and while he was honest about the provenance of his artefacts, others can be, shall we say, less scrupulous.

The issue of selling the artefacts, which is still forbidden, will have to be dealt with sooner or later. It's going to be difficult for a private company, whose profits come mainly from the sale of exhibition tickets, to preserve and maintain the relics of the *Titanic* forever. But I believe that their dispersal at spectacular auctions should be avoided.

As we have many iterations of the same object, the solution seems to be to build up several collections. These could then be sold or transferred to maritime museums with a special interest in the *Titanic*, such as the Cité de la Mer in Cherbourg or the Titanic Belfast Experience in Ireland.

Chapter 4

May–June 1993

A gold watch and the clandestine clarinettist's suitcase

In spring 1993, a new expedition was launched. It brought together the Americans from RMS Titanic, Inc. – the new name for Titanic Ventures – and the French staff of Ifremer. Fifteen dives were made in fifteen days, including four with the *Robin* (the Nautile's inspection robot) into the three forward holds, the starboard forward breach, the grand staircase and the radio control room. Exploration depended on the *Nautile*, the *Robin* and the good old *Nadir* as the support vessel. Based on our experiences in 1987, we drew up a precise programme of dives. The campaign summary could not have been clearer:

This campaign was the subject of contract 93/1211861/F signed on 7th May with RMS Titanic, Inc. The duration of its contract was 42 days, with the possibility of a 10-day extension. Departure from Toulon on 27th May, arrival in Concarneau on 8th July.

The Nautile's dives had the following objectives:

☐ *a detailed inspection of the front end, in particular the three holds*
☐ *the starboard breach*
☐ *the Grand Staircase*
☐ *the radio control room*
☐ *to bring up objects, some of them very large, from the debris area.*

A programme that we obviously did not completely stick to. Either the conditions were not as we had envisaged, or we didn't get to the planned spot, or we were looking for something we couldn't find, and vice versa . . . Unlike scientific diving, where you know exactly what you're going to do, how and why, diving on the *Titanic* means flirting with the unexpected. That said, the surface often reminded us that we needed to stop dawdling. The mission came first.

Six years earlier, we had brought up some egg dishes. This time, we were determined to get the whole set – no fewer than 240 items. The idea was to show them at exhibitions, faithfully reproducing the strange way the dishes had lain on the ocean floor. The kitchen cupboard that held them had vanished, but on the seabed, the dishes were neatly arranged as they had been originally. The set was to be displayed on a bed of sand for added realism. Today, visitors to *Titanic* exhibitions find it to be one of the most spectacular and evocative items on display.

On 21st June, we also recovered a complete davit[1]. Built in London by Welin Davit & Engineering Co., Ltd[14][15], this quadrant model, new for its time, weighs in at a whopping 1.6 tonnes.

We all had another, unwritten objective in mind: a repeat of the stroke of luck we had in 1987 when searching the debris field for bags and luggage. Most of the suitcases we brought up were empty, probably because passengers had packed their belongings in the cabin cupboards at the start of the journey. No jewellery or precious stones this year. But we did discover two objects of priceless sentimental value: a gold watch and an ordinary, used suitcase.

The pocket watch lay isolated on the sediment. The *Nautile* had seen it after setting down to retrieve another object. This was pure luck. Unless caught in the glare

---

[14] Lifting device on board a ship, for lowering lifeboats.
[15] This specialist company still exists today, under the name Welin Lambie.

of a searchlight, such a small object is difficult to see, given that our submersible travels some 2 metres above the seabed.

As was customary at the time, the initials of its owner were engraved on the case. A perusal of the passenger list revealed that it belonged to Thomas William Solomon Brown, a hotelier in South Africa. Aged sixty, he was travelling with his wife, Elizabeth Catherine, and daughter Edith Eileen, then aged sixteen. With business becoming difficult in South Africa, the Brown family had decided to leave the country and settle in Seattle, USA, where Thomas's sister-in-law lived. Their plan was to open a hotel there. They took with them crockery, furniture and a thousand linen sheets purchased during a long stopover in England. Thomas Brown died in the shipwreck, but Elizabeth and Edith survived. We found out the latter was alive and living in England – Southampton to be precise – the place from which the *Titanic* had set off on its fatal voyage.

George Tulloch, President of RMS Titanic, Inc, decided to have the watch restored and returned to Edith, aged ninety-seven. Entrusted to Jerry Sussman, a watchmaker in Fairfield, Connecticut, it underwent a discreet cleaning that enhanced the finely chiselled gold case, apparently sourced from France. As for the rest of the watch, while the dial held up fairly well, the highly-oxidized hands are permanently fixed at 11.04 (or 23:04). The mechanism was a mass of corroded metal. As Sussman explained, 'It's very emotional having this watch in my hands. From a watch-making point of view, I'd say it's worth peanuts, but from a historical point of view, I'd estimate it at $250,000.'

Tulloch, who was very skilled with his hands, made a wooden display case with a plaque bearing the inscription: WHAT BETTER USE OF SCIENTIFIC TECHNOLOGY THAN TO REUNITE A FATHER WITH HIS CHILD? The watch was presented to Edith Brown (Edith Haisman since her marriage), in a moving ceremony in a Southampton hotel. When Edith set eyes on it, she was transported back eighty-one years to the precise moment when, safe at last with her mother in lifeboat No. 14, launched at 1.30am, she saw her father for the last time. Still on board the liner, Thomas shouted, 'I'll see you in New York!' and automatically pulled his watch from his vest and looked at the time . . .

A special dispensation allowed Edith Haisman to keep the watch for the rest of her life. It would return to the collection of artefacts recovered from the *Titanic* on her death. Edith died on 26th January 1997, months after receiving a telegram from Bill Clinton congratulating her on her 100th birthday on 27th October, 1996. She received a second telegram half an hour later, this time from Queen Elizabeth II. She was an energetic, mischievous woman who once asked me, 'Mom left a necklace on her cabin nightstand, would you mind retrieving it?' Unlike Eva Hart, she considered it a good thing to find and bring back *Titanic* artefacts like her father's watch.

The old suitcase told an even more extraordinary story. When I spotted it through the *Nautile*'s porthole, my eye was drawn to a score. I can't read music, but I recognized staves and, underneath, English lyrics. Brought on board, it was entrusted to the conservation laboratory to stabilize and render the papers it contained legible. The suitcase didn't have much to reveal, except scores of popular songs and extracts from musicals, such as the serenade 'The Land of Romance' from *The Belle of the West*, by Karl Hoschna, a fashionable Broadway composer. There was also a clarinet, letters well enough preserved to still be legible, a notebook which had been used as a diary, playing cards, etc. Everything pointed to it being the luggage of a musician from the *Titanic*'s orchestra, a certain Howard Irwin, if the envelopes were anything to go by.

Except there was no Howard Irwin in the orchestra. And his name was nowhere to be found on the list of passengers who boarded the *Titanic*. From then on, the most far-fetched hypotheses were rife among us: Irwin was a stowaway – but stowaways rarely carry luggage. Irwin was the pseudonym of a disreputable individual, a professional gambler (and probably cheat) who operated all over the planet, as evidenced by the card game and horse and greyhound race tickets from Australia . . . Irwin was the man disguised as a woman who Edith Haisman claims to have seen jumping aboard lifeboat No. 14, where she was, and who disappeared once safely on the *Carpathia*[16]. Above all, Howard Irwin was elusive. He was a ghost, yet his suitcase was real, as were the letters we found, sent by his fiancée, Pearl Shuttle.

The enigma could have remained complete if it hadn't been for a chance occurrence. On 20th April 1997, Dave Shuttle and his wife watched a Discovery Channel documentary, *Titanic: Anatomy of a Disaster*. Suddenly, Dave Shuttle froze: a letter, signed 'Mrs Shuttle' and clearly alluding to his Great Aunt Pearl, appeared on the screen. It was a letter written by Dave's great-grandmother to Howard Irwin to reassure him of his fiancée Pearl's fondness for him. According to the letters and diary found in the suitcase, a jealous Howard was very concerned about Pearl's fidelity. She was a ravishing young woman who suited her name – she was a musician on tour with her orchestra across the United States. Pearl was finding it hard to tolerate the interminable voyage Howard had embarked on. The eighteen letters found in the suitcase

---

[16] See James Pellow and Dorothy Kendle, *A Lifetime on the Titanic: the biography of Edith Haisman*, London, Island Books, 1995.

salvaged from the wreck tell the story of their love affair by correspondence, a love that disintegrated over time and distance, culminating in a final seven-page letter in which Pearl hinted ('I'm so tired of waiting . . .') at her desire to put an end to their relationship. From that moment on, events snowballed. Dave Shuttle managed to make contact with Howard Irwin's family, and the unbelievable truth emerged. Yes, Howard's suitcase did board the *Titanic*, but without him.

Howard Irwin was born on 13th December 1887 in Lindsay, Ontario, Canada. He crossed the border to work in Buffalo in the fledgling automobile industry. In early 1910, he decided to embark on a round-the-world trip with his friend Henry (Harry) Sutherland, who worked in the same factory. The two rascals relied on their musical talents to make a living. Howard played the clarinet, Harry the violin. After a long stay in Australia, they split up to travel separately – Irwin to Africa, Harry to the Far East – and planned to meet up again in London. This they did in early April 1912, to return to the United States. Their liner had to remain docked due to a shortage of coal caused by a long strike by British miners. They were transferred to the *Titanic*, as the White Star Line had requisitioned all available fuel to ensure the maiden voyage went ahead. Nobody says no to travelling on the world's most prestigious ocean liner, even in third class! On 9th April, the eve of departure, the level-headed Harry decided to go to bed early. Howard, who was somewhat livelier, paid a visit to the pubs in Southampton harbour. From pub to pub and from pint to pint, the night progressed, tempers flared, and Howard, already completely drunk, got caught up in a brawl. He took a blow to the head and all went black.

On the morning of 10th April, noting the absence of his travelling companion, Harry decided to board alone, but took on Howard's suitcase in the hope that he would soon arrive at the dock. When the *Titanic* sailed, there was still no sign of Howard because at that very moment, Howard was waking miserably from his restless night to discover that he, too, was on board a ship. But this vessel was no luxury liner. While unconscious, in the grand tradition of the port underworld, he'd been 'shanghaied', i.e. kidnapped, and forcibly conscripted into the crew of a rotting cargo ship bound for China. On 15th April, news of the sinking of the *Titanic* reached him. He decided to return to the United States at the earliest opportunity. During the Port-Saïd stopover, he escaped from his prison barge, found a place as a sailor on a ship bound for England, and disembarked in Southampton virtually penniless.

Ever inventive, Howard tried his luck as a stowaway on the *Olympic*. For two days, he hid in the boiler room amid the suffocating heat and toxic fumes. Once the liner was in international waters, he popped up black with coal, and starving. He turned himself in and signed on as a crew member to pay for his crossing.

When he reached Buffalo a month later, Howard learnt that Harry had been one of the victims of the disaster. His wonderful Pearl had died of pneumonia at the end of the previous year. He had nothing left. From then on, he opted for a more settled life. In 1914, he married Ivy Curiston, a young girl from Michigan who was fascinated by this romantic character. Their descendants say Howard and Ivy were very happy together.

But that didn't stop Howard Irwin being seized once again by the demon of adventure. In 1923, he took part in the Klondike Gold Rush in Alaska. He returned richer . . . in experience. The reformed globetrotter eventually became a foreman at the Curtiss-Wright Corporation, a New Jersey aircraft plant. He died in 1953 without ever having lost his wanderlust. Fascinated by R. L. Stevenson's novel *Treasure Island* since childhood, he kept a map he had found in Tahiti which he believed had been drawn by pirates, until his dying day. And he dreamt of going there to find the fabulous booty.

Howard Irwin's family has preserved a wealth of documents about his life, some in his own handwriting, and numerous photos. It's thanks to these archives that we have been able to reconstruct the essence of his extraordinary itinerary. Looking through his diary, I came across a sentence that seemed troubling in light of his odyssey. 'Beer brings joy, don't forget that water only makes us wet.'

Around that time, George Tulloch told me there was one last French survivor, Michel Navratil. He asked me to get in touch with him, as I still lived in France and was the only French speaker at RMS Titanic. So, I investigated. Michel Navratil was living in Montpellier, where he had retired after a career as an associate professor of philosophy and then chair of psychology at the city's Faculty of Arts. Having got hold of his telephone number, I called him. His courteous reply was less than encouraging.

— Look, I don't know if I like what you're doing or not, so I'll have to think about it. I'm in the process of tidying up my library, so you can call me back later.

— When?

— Give me three months.

As soon as three months had passed, I called him back. This time, he was up for it. I jumped in my car and headed for Montpellier. And on this first meeting, we spent the whole day together. For twelve hours, the former teacher talked about the *Titanic*, how the sinking had influenced his entire existence, made him think about the problem of death at an early age and turned him towards philosophy. I got a real philosophy lesson, and an exciting one at that!

Michel Navratil's destiny was out of the ordinary, even for a *Titanic* survivor. While most of those rescued were reunited with their families within days, Michel, not yet four, and his little brother Edmond, just two, were the only surviving children not claimed by a parent. Edmond could only babble, Michel was just able to say, 'I'm Lolo, he's Monmon,' and as he said it in French, nobody understood. The identity of the '*Titanic* orphans', as the English-speaking press dubbed them, remained an enigma for many long weeks.

It has to be said, their father did everything in his power to cover his tracks when boarding the *Titanic*, to make sure it was impossible to trace them. Of Slovakian origin, Michel Navratil (he gave his first name to his eldest son) was a tailor in Nice. Business had been good, but a few months after Edmond's birth, he discovered his wife, Marcelle, was having an affair with a baron – a cavalry officer in the Italian army. Deeply wounded, he filed for divorce and then, in early 1912, decided to start a new life and moved to Chicago, where some of his family had emigrated. The tailor then sold his store to a colleague, Louis Hoffmann. The second part of his plan was now set. At Easter, while visiting his children, he kidnapped them and left for England. He embarked at Southampton rather than Cherbourg, where customs officials might have been suspicious. To better cover his tracks, he had bought second-class tickets in Monte Carlo and borrowed Louis Hoffmann's passport to travel under a pseudonym. So, it was the Hoffmann family that boarded in Southampton – the father, claiming to be a widower, and his two young children. Once in the US, it seems the tailor intended to ask his ex-wife to join him and start their lives again from scratch. He hoped the prospect of being reunited with her boys would convince her to join them.

When the shipwreck occurred, the father put his children in the collapsible lifeboat D, the last to leave. Michel remembered his father's last words perfectly, tender lines for his mother that he instructed him to repeat ('Tell your mother I send her all my love . . .'). There was also the impact of the lifeboat hitting the water before drifting away into the dark night – a very distinctive sound, engraved in his memory. The two brothers then fell into a deep sleep. Michel woke at dawn, just as he was about to be transferred to the *Carpathia* in a canvas bag, like a parcel; a manoeuvre he still has bad memories of. One of the passengers in lifeboat D, Margaret Hays, took charge of these unaccompanied children.

When they arrived in New York, journalists were fascinated by their story. The 'orphans of the Titanic', if not the 'orphans of the deep', made headlines. While waiting to be officially identified, Michel and Edmond were placed by Miss Hays with a family she knew, the Tylers. They lived on a large estate near Philadelphia and had children of roughly the same age. A friendship was born between Michel Navratil and Sidney Tyler. The two men remained close friends until Sidney's death in 1993.

In Nice, Marcelle, bereft of her children, was worried sick. Never in a million years would she have guessed the truth. It was while reading an article in *Nice-Matin* on 21st April 1912, that she discovered that two young French children who had survived the sinking of the *Titanic* were waiting for their mother in America. Seized by doubt, she hurried to the British consulate in Nice, who contacted the French consulate in New York. After exchanging dispatches and photographs, her suspicions were proven. Lolo et Monmon really were Michel and Edmond Navratil. Marcelle embarked on the *Oceanic*, a transatlantic liner of the White Star Line (it seems the company had gifted her the round trip with the children) and was reunited with her sons in New York on 16th May 1912. All three returned to France. The body of Michel Navratil senior, initially reported missing, was pulled from the sea around 20th May by the *Mackay-Bennett*. Among his personal belongings was a passport in the name of Louis Hoffmann. Presumed Jewish because of his surname, he was buried in the Baron-de-Hirsch Israelite cemetery in Halifax. It would be two long years before the Navratil family obtained official recognition that Louis Hoffmann was in fact Michel Navratil and were able to have his name inscribed on his tombstone.

Michel Navratil Jr. died on 30th January 2001, aged almost ninety-three. He was the last male survivor of the shipwreck, and the last French person, too. I went to his funeral and keep in touch with his family, especially his daughter Élisabeth. She's been fascinated by the story of the 'Titanic orphans' since she was a teenager, as it is also her story. She wrote a book about her father and Uncle Edmond's adventure, Les Enfants du Titanic[17] (The Children of the Titanic), and created an opera, Mozart Titanic. Her concern for truth and accuracy led her to revise and add to her book several times as she discovered new material. I salute her tireless determination to pass on the memory of the sinking and the victims through first-hand accounts. They act as an antidote to the endlessly copied and recopied myths that tend to reduce the story of the Titanic to pure fiction.

---

[17] Élisabeth Navratil, Les Enfants du Titanic, Paris, Hachette, 2012.

Chapter 5

Summer 1994

Coal chores!

'To summarize, the Nautile has: made eighteen dives to the site; spent 184 hours diving, including 114 hours at the bottom; recovered 764 objects, including one weighing 2,350kg and one weighing 1,200kg; raised 170 pieces of coal, i.e. 2,340kg; recorded 220 hours of video.'

The expedition report reflects our objective, which was unusual to say the least, and probably unique in the annals of underwater exploration, namely: to raise some of the *Titanic*'s coal. These large blocks of anthracite weighing up to 50 kilos – the stokers used to break them up with crowbars before they were fed into the boilers – lie in abundance throughout the debris field. It is estimated that, of the 5,892 tonnes loaded in Southampton, some

2,800 tonnes remained in the bunkers at the time of the sinking, given a consumption rate of 600 to 850 tonnes per day. Less prestigious than diamonds, but a precious testimony to the era of the great steamers, these blocks have their rightful place in *Titanic* exhibitions.

During the course of the mission, RMS Titanic, Inc. decided to intensify the harvest and put most of it on the market to finance future expeditions. The ban on selling *Titanic* relics actually only applied to manufactured items – not coal, which is considered a mineral.

The 170 blocks of coal were broken up into 400,000 small pieces and offered for sale in the autumn of 1995 at a price of $25 (the equivalent of $45 today). They came in a display case with a certificate of authenticity signed by both the President of RMS Titanic, Inc. and myself. This coal had been physically present on board the liner and was released from the hull when it broke in two, only to be brought to light when the wreck was being explored. In its own way, it's a silent witness to history. We could have sold entire blocks at auction, and they would undoubtedly have fetched huge sums. But we chose to make these authentic relics available to everyone.

Harvesting the coal led us to adapt the *Nautile* for use as though the debris field was a construction site. As it was out of the question to go back and forth to the nearest basket for each piece of coal, or to use the submarine's scientific basket, we had the *Nautile* fitted with a hydraulic tipping bucket. It was, as always, built on board thanks to the indefatigable container workshop and the no less inexhaustible resources of Christian Le Guern and his team (Pierre Valdy was not with us that year). When the bucket was full, we simply tipped it into a basket – a recovery operation that we privately dubbed 'coal chores'.

Titanic coal is doubly interesting, from both a technical and a legal point of view. On the technical side, it's worth noting that while the liner was intended to be state-of-the-art and fitted with watertight bulkheads remotely controlled from the bridge, on-board telephones, Marconi wireless, electric cranes, several batteries of interior elevators, etc. she was one of the last new transatlantic liners to retain the tried-and-tested, but already outdated, solution of coal-fired heating. This was a time when oil-fired heating was beginning to gain ground, notably on US Navy ships. Unlike the oil-rich United States, Great Britain had considerable reserves of high-quality coal, which justified this conservatism. Yet coal's disadvantages were glaring, especially on a giant of the seas whose consumption was proportionate to its size. No fewer than seventy stokers shuttled between the coal bunkers and the boilers with wheelbarrows, and 170 boiler men toiled in the din and heat of 50–60°C to feed the huge machines. Each was fitted with a buzzer to warn that it was time to refill the coal tank. On average, it went off every seven minutes. With twenty-nine boilers to feed twenty-four hours a day, the men were forced to slave away like convicts. This workforce was basically housed in a dormitory at the very front of the ship, well below the third-class berths and directly linked to the boiler room by a special corridor. This tunnel in the bottom of the hull ensured that these 'black faces', as they were nicknamed, would not frighten passing passengers. It all seems very archaic . . . And by 1919, the *Olympic*, the only survivor of the series, would be modified to run on oil-fired boilers.

From a legal point of view, the exceptional status of the coal, whose sale was legal, sums up the complex status of the wreck. Which bits are part of the hull? Which items are the passengers' own property? What belongs to third parties who entrusted a cargo to the company? The main principles of the insurance code and maritime law shed light on the situation.

Initially, after the sinking, insurance came into play, and the *Titanic*'s co-owners, the White Star Line and the International Mercantile Marine Company, were indemnified by their consortium of insurers, who automatically became the owners of the wreck. Passengers or their families were compensated by their own insurance companies. However, some families received nothing, as there was no list or inventory proving that a relative who was shipwrecked had possessions on board. On the other hand, the contents of the ship were not subject to the same regime as the hull. For example, the 3,000 or so bags of mail still in the sorting room and postal hold remained the responsibility of the Royal Mail until their contents were delivered[18]. But the same cannot be said of lost registered mail, for which the US government claimed and obtained lump-sum compensation.

Later on, the *Titanic*'s insurers effectively abandoned the wreck: no salvage action was attempted or even anticipated. The liner's hull became a 'thing in dereliction' (*res derelictae* for jurists), no longer legally owned.

The *Titanic* lies in international waters, which means that no country has *a priori* rights over the wreck. Only military vessels or those built and financed by a State remain the property of that State, wherever they may be, even in the territorial waters of another country. The *Titanic*, a commercial vessel owned by a private company, is obviously not covered by this clause.

There is, however, an international maritime law, the Admiralty Law, which stipulates that anyone who finds a wreck and wishes to obtain rights over it must apply to the competent authority. This varies from country to

---

[18] The *Titanic* was referred to as an 'RMS' (Royal Mail Ship) rather than an 'HMS' (His/Her Majesty's Ship) because she carried Royal Mail letters and parcels under contract.

country, depending on the nationality of the discoverer: a federal judge in the United States; the Receiver of Wreck in Great Britain; or a Maritime Affairs Administrator in France. In support of his request, the finder must indicate the exact position of the wreck and present an item recovered from it. There is a waiting period that varies from country to country but is at least one year and one day to enable the authorities to check whether there is a legal owner. Once this has elapsed, if their application is accepted, the discoverer (or inventor), becomes salvor-in-possession and holder of salvage rights over the wreck.

Strangely, until 1992, no one had bothered to take the official steps to become a salvor in possession of the *Titanic* wreck – neither the Woods Hole Oceanographic Institute nor Ifremer. Robert Ballard, though keenly interested, was unable to do so for one simple reason. The law stipulated that an item from the wreck had to be produced for the request to be accepted, and he was forbidden by the director of the Woods Hole Oceanography Institution to take anything from the *Titanic*. To be a beneficiary of this law, he would have been obliged to put himself outside it. Perhaps he hadn't thought of it, just like Ifremer. The situation was made even more intractable by the fact that Ballard himself had taken the lead in opposing any recovery. With Ballard out of the picture, Jack Grimm, who had not given up his claim to the wreck, felt he had a card to play.

On 7th August 1992, the Texan billionaire applied to the federal court in Norfolk, Virginia, for Marex Titanic to be recognized as salvor-in-possession, and attached to his application the requested proof: a small glass vial he claimed was from the *Titanic*. Knowing Jack Grimm's reputation, Judge J. Calvitt Clarke Jr., who was in charge of the case, investigated the joint Ifremer/Titanic Ventures expedition of 1987. He could find no trace of Marex Titanic at that time. Grimm persisted, making more and more demands. Not only did he want to be recognized as the salvor-in-possession, he was also demanding the return of the 1,852 artefacts brought up by Ifremer. He cited his anteriority, claiming he had discovered the wreck as early as 1981. Titanic Ventures struck back as testimonials multiplied (the transcript fills 425 pages). One day, the judge was questioning Marex's representative.

— Where did the bottle you say was recovered from the wreck of the *Titanic* come from?

— It was given to me during the 1987 expedition.

— By whom?

— Paul-Henri Nargeolet!

Called to the stand, I issued a formal denial and I would later write a sworn statement. To avoid this type of incident, all items under my supervision on board the *Nadir* were kept in a container locked with two padlocks. The captain had the key to one, and I to the other. Two people were needed to open the container and remove items. I also explained that some of the items were taken aboard the second expedition support boat, over which I had no control. And it's highly likely that this was where the bottle was filched. The judge took a dim view of Marex's claims, being based on a stolen object . . .

Despite a number of tortuous manoeuvres, Jack Grimm and Marex Titanic were unsuccessful in a legal battle that made history in the annals of American law. Grimm disappeared from the scene, and Titanic Ventures' application to be recognized as salvor-in-possession (in order to cut short any further legal action) followed its course until 7th June 1994, when the Norfolk court confirmed this status. This recognition carried the following condition, proposed by Titanic Ventures, in this case by its president, George Tulloch. The artefacts brought up must remain effectively *in possession*, which means they cannot be dispersed; only the sale of the entire collection can be envisaged. This implied that the company's mission was to generate revenue by exhibiting *Titanic* artefacts, not by selling them. Titanic Ventures, Inc. then changed its name to RMS Titanic, Inc., still managed by George Tulloch. On the strength of the 1987 experience and the discovery of the bag full of jewels, the company ended up buying back all rights to passengers' personal belongings from the insurance companies, a precaution that would head off future controversy. In exchange, RMS Titanic, Inc. also transferred some items from the 1987 expedition to the insurers, who in turn transferred them to the Merseyside Maritime Museum in Liverpool, where the *Titanic* was registered[19].

The situation that the wreck and the artefacts brought up seemed to have been settled once and for all. However, it became apparent that this was not the case. The *Titanic*'s legal tribulations were far from over. Three episodes in particular illustrate the complexity of the law in this area.

The first concerns the request made in 2002 by the new management team of RMS Titanic, Inc. was to be allowed to disperse some of the recovered items. This request was rejected because the ruling of 7th June, 1994 applied to RMS Titanic and not to its directors, so the change in the company's management did not justify amending its terms. In court, the exchange between the plaintiff and

---

[19] The same museum also exhibited the famous silver cup given to the captain of the *Carpathia*, Arthur Rostron, by 'the unsinkable Molly Brown', a survivor of the *Titanic*. The trophy was sold at auction in 2015 to a private collector for $200,000.

Federal Judge Rebecca Beach Smith, who had assisted Judge Calvitt Clarke in the 1994 decision and succeeded him on his retirement, was brief:

— We request permission to sell the items individually.

— In 1994, your company specifically asked for artefacts to be protected against individual sale.

— That's right, but we've changed our minds.

— Maybe, but I haven't changed my mind!

This just goes to show the extent to which a federal judge is sovereign in his or her interpretation of the law, regardless of the arguments put forward by the defendants. Her refusal, in this case, preserved the unity of the collection.

The second episode focuses on the reassessment of RMS Titanic's status. In 2009, it asked to be confirmed as the owner of the *Titanic*'s artefacts, thus conferring the right to sell them. The company argued that this sale, if authorized, would be a bulk sale to a museum. Judge Smith ruled that RMS Titanic could not be considered the owner of the artefacts, only their custodian and rescuer. This change in status had a direct and immediate consequence: the case moved from the property register to that of sea rescue. Maritime law is crystal clear on this point: the salvor is entitled to compensation commensurate with the effort expended in carrying out the rescue. The compensation to be paid to RMS Titanic was calculated at $225 million . . .

The *Titanic* was built by Harland & Wolff in Belfast between
1910 and 1911. The ship sank on 15 April 1912 after colliding
with an iceberg on its maiden voyage to New York.

One of the *Titanic*'s magnificent first-class lounges,
with intricately-carved oak panelling.

A rare photograph taken on the day the *Titanic* was launched, 31 May 1911.

© Harris Museum and Art Gallery/Bridgeman Images.

From left to right, First Lieutenant Murdoch, Second Captain Wilde, an unidentified officer and Captain Smith, who did not survive the sinking of the *Titanic*. Photo taken aboard the *RMS Olympic*.

© CSU Archives/Everett Collection/ Bridgeman Images.

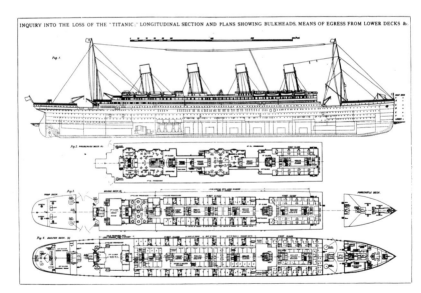

From top to bottom: view of the starboard side, promenade deck A, deck B (forecastle, bridge deck, sterncastle) and deck C (shelter deck).

© The Stapleton Collection/Bridgeman Images.

Launch of the *Nautile* by the *Nadir* at the *Titanic* site.
The submarine can dive to a depth of 6,000 metres.
© Christian Petron Cinemarine.

The bow of the *Titanic*. Photo taken by the *Titan* submersible.
© OceanGate Expeditions, Ltd.

One of the four cylinders of the *Titanic*'s reciprocating machine. The machine equals four storeys in height. Photo taken by the *Nautile*.
© RMST, Inc.

Captain Smith's bathtub, full of debris from the ceiling of his bathroom. He had fresh water and hot and cold seawater on tap. Photo taken by the *Nautile*.
© RMST, Inc.

View of the *Titan* five-seater submersible approaching its launch and recovery platform.
© OceanGate Expeditions, Ltd.

Binoculars displayed in an aquarium before restoration. Several pairs of binoculars found on the wreck had been made in Paris.
© Paul-Henri Nargeolet.

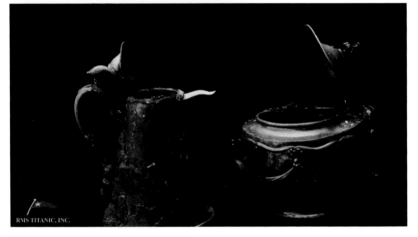

Solid silver chocolate pots before and after restoration. They were standing on iron plates and were coated with iron oxides that were removed by electrolysis. © RMST, Inc.

Leather travel bag, the handles of which had been cut with a sharp tool, containing numerous jewellery cases, gold coins and banknotes.
© RMST, Inc.

Watch belonging to Mr Brown, who died in the shipwreck. It was given to his daughter, Edith Haisman, who survived.
© RMST, Inc.

The 3rd-class purser's safe. This fire-resistant safe, found in the debris field near the stern, contained a bag of coins and some documents. Photo taken by the *Nautile*.

These 240 egg dishes were once stored in a wooden cabinet. It has disappeared and the dishes were found in this arrangement on the sediment. All were brought to the surface. Photo taken by the *Nautile*.

Melvina Dean, the youngest passenger on the *Titanic*, only a few weeks old at the time of the sinking, became its last survivor. Here she is with the author, at the arrival of the *Royal Majestic* – one of the cruise ships that took part in the 1996 expedition.

RMS Titanic Expedition 1996 © Xavier Desmier/Gamma Rapho.

A young boy exhibiting the excellent book *Titanic: Triumph and Tragedy* (Norton & Company, Incorporated, W.W., 1995) by historians Charles Haas and John P. Eaton, who dived to the wreck with the *Nautile* and took part in several expeditions.

RMS Titanic Expedition 1996 © Xavier Desmier/Gamma Rapho.

Ceremony on board the *Royal Majestic* at the site of the shipwreck,
to commemorate the passengers.
RMS Titanic Expedition 1996 © Xavier Desmier/Gamma Rapho.

Edith Haisman, a *Titanic* survivor whose father perished with
the ship, attending the ceremony with two other survivors,
Michel Navratil and Eleanor Shuman.
RMS Titanic Expedition 1996 © Xavier Desmier/Gamma Rapho.

A toast during the interview with Michel Navratil, one of the survivors of the shipwreck. He and his brother Edmond were nicknamed the '*Titanic* orphans'. To his right, his daughter Elisabeth Navratil, and behind them Michele Marsh, presenter and reporter at NBC, and the author. © Michele Marsh.

A moment of joy for reporter Sarah James and the author after the successful recovery of the 'Big Piece', a 20-tonne piece of the *Titanic*'s hull, aboard the *Abeille Supporter* in 1998.

Personal photo from the author's collection.

George Tulloch, Paul-Henri Nargeolet and Pierre Valdy a few minutes after the Big Piece was hoisted onto the *Abeille Supporter*.
© Günter Bäbler.

The *Abeille Supporter* positions itself to retrieve the Big Piece. Two diesel bladders (BAGs) are visible on the surface. The Big Piece is suspended with other BAGs below the surface.
© Christian Petron Cinemarine.

Ceremony marking the arrival of the Big Piece in Boston on the
Abeille Supporter. © Christian Petron Cinemarine.

Hoisting the Big Piece aboard the support vessel.
The author, on the right, supervises the operation.
© Christian Petron Cinemarine.

The astronaut Buzz Aldrin and the author on board the *Nautile*.
RMS Titanic Expedition 1996 © Xavier Desmier/Gamma Rapho.

On board the *Royal Majestic* with, from right to left, Mrs Aldrin,
Buzz Aldrin, George Tulloch and the author. © Günter Bäbler.

On board the *Ocean Voyager*, biologist Roy Cullimore examines
a fragment of an iron oxide stalactite containing bacteria.
RMS Titanic Expedition 1996 © Xavier Desmier/Gamma Rapho.

Guy Sciarrone,
engineer and chief
pilot of the *Nautile*,
with one of the
Titanic's whistles, on
the deck of the *Nadir*.
© Eric Lacoupelle.

Conservationist and restorer from LP3 Conservation taking notes on
one of the solid silver tureens recovered from the debris field.
RMS Titanic Expedition 1996 © Xavier Desmier/Gamma Rapho.

On reaching the surface, each object is catalogued,
measured, photographed and documented.
RMS Titanic Expedition 1996 © Xavier Desmier/Gamma Rapho.

A reunion on board the *Nadir* of its crew, teams from the *Nautile*,
Ifremer, Aqua+, NBC, LP3 Conservations, RMS Titanic, Inc., historians,
photographers and cameramen involved in the 1996 expedition.

As no official body was in a position to pay such a sum, the affair reached an impasse and was quietly forgotten. RMS Titanic was confirmed as owner two years later. Since then, the public has been convinced that the collection of *Titanic* artefacts is 'worth' $225 million. In reality, the sum bears no relation to the value of the objects, which remains unknown. Some of them would have to be sold to get an idea of their market value, and that's impossible. The collection could be worth much more . . . or much less.

The third and final twist (to date) was the permission granted in 2020 by the same Judge Rebecca Smith to RMS Titanic to raise the roof of the Marconi Room[20] in order to recover the *Titanic*'s wireless equipment. In principle, nothing was to be taken from the wreck itself. But it was beginning to subside, and there was an urgent need to preserve these devices, which are of major historical and technical interest. It was thanks to them that the other ships were alerted, and they were the newest and most powerful devices of their time.

The Marconi Room was not on a deck, but on the outside of the *Titanic*, at the very top of the ship. And yet, the very

---

[20] The name given to the ship's radio room, with equipment and operators supplied by Marconi's Wireless Telegraph Company, a British company founded in 1897 by the physicist Guglielmo Marconi (1874-1937), inventor of wireless telegraphy.

idea of touching even a sheet of metal to free the devices caused a scandal. The old argument of violating the burial grounds of 'the remains of 1,500 people' resurfaced, this time brandished by the National Oceanic and Atmospheric Administration (NOAA). This American agency claimed to represent the 'public interest' in this affair, and its links with Robert Ballard are long-standing and well known. This new media campaign did not change the judge's decision. In her view, the recovery operation was not going to be carried out 'in' the wreck, but 'on' it. Unfortunately, the Covid pandemic forced us to cancel the expedition scheduled for summer 2021.

All these legal twists and turns speak volumes about the passions unleashed by the *Titanic* and its legend. But make no mistake: they were merely a storm in a teacup. American court rulings, press campaigns and multilateral conventions be damned. The wreck of the *Titanic* was well and truly in international waters. In other words, nothing and no one could prevent a country that wasn't a signatory to these conventions from launching an expedition and using the wreck as it saw fit. America's decisions apply only to its own citizens. The wreck preservation agreement between the United States and the United Kingdom binds only these two countries and their nationals, who must apply for authorisation to undertake work on the *Titanic* – but not to dive down to it. Not to mention that this agreement contradicts the US Constitution, which states that one person cannot be prevented from undertaking work if another person has been authorized to do so . . . In short, this agreement prevents American and British citizens from doing what citizens of other countries can freely do. To be truly effective, such an agreement would have to be signed at UN level.

As far as I know, no one is planning a major operation on the *Titanic* at the moment, but some countries have the means to do so. Recently, I went to Shanghai, where I met specialists in sea rescue. I was able to see the type of equipment they have at their disposal – colossal boats which are unparalleled anywhere in the world. They are working on tools that can open up wrecks to recover their contents. If they decide to go to the *Titanic* and extract the strongroom, they'll succeed. There would be a public outcry, but it wouldn't change a thing. Yes, in 2012, under pressure from NOAA, UNESCO declared the *Titanic* a 'world heritage site'. But in my opinion, this makes no sense, because it's a heritage site destined to disappear, so this protection of underwater cultural heritage is illusory. What happened with the destruction of the giant Buddhas and statues in Afghanistan could happen here. An avalanche of energetic condemnations and vigorous protests, but no one lifting a finger.

The wreck also faces other natural threats, such as underwater earthquakes and boulders released by icebergs. It was long believed that the 1929 Newfoundland submarine landslide, triggered by a major earthquake, had buried the wreck of the *Titanic*. This was not the case, but it could still happen. Boulders released by melting icebergs abound on the site and come in all sizes, from large pebbles to huge blocks of stone. One of them, weighing around 50 tonnes, fell just 50 metres from the front of the wreck. The impact, after a fall of almost 4 kilometres, came within a hair's breadth of crushing everything.

Under these conditions, it's a miracle that the wreck is still in relatively decent nick. Instead of finding flimsy reasons for putting the brakes on exploration, it would be better to speed it up while it's still feasible. That way, we can learn as much as we can and protect as many relics as possible by bringing them to the surface, even if this means moving a few rusty metal sheets. Future generations are likely to thank us.

Chapter 6

July–September 1996

Five breaches and a bouquet for a shipwreck

'1996 RMST, Inc.—Ifremer—Discovery Channel—NBC
M/V Nadir + Nautile + Robin, M/V (or RV) Ocean Voyager
(studio) and M/V Jim Kilabuk. 8/4 to 9/1/96.
Exploration, recovery. Use of lighting towers. Attempt to recover
the Big Piece (BP). The BP returns to the bottom of the ocean,
10 NM away from the Titanic site.'

The telegraphic style of this campaign report requires a little deciphering. M/V stands for 'motor vessel'. The *Ocean Voyager* is a research vessel used on this occasion as a studio for the production of a documentary for the Discovery Channel, NBC and Canal+[21]. The *Jim Kilabuk* is

---

21 *Titanic, anatomie d'un naufrage (Titanic, anatomy of a shipwreck)* directed by Sylvain Pascaud for Canal +/Discovery Channel, 1996.

an assistance vessel. The four 'Edison' lighting towers, each equipped with five 1,200-watt spotlights, were used to illuminate the wreck during the shoot. As for the recovery of the BP, the Big Piece – a monumental 20-tonne fragment from the *Titanic*'s hull – it gave us a hard time that year. '10 NM' represents nautical miles, or nearly 20 kilometres.

As is often the case, this summary misses the key point: the presence of engineer Paul Mathias on board the *Nadir*. He was in charge of equipment we were using for the first time on the wreck, a very low-frequency (3 kilohertz) sediment-penetrating acoustic sounder designed for geological research. This type of device can penetrate the ground at depths of up to 160–180 metres and take images of sedimentary strata and buried rocks. The images are of medium definition and cover a fairly large area. We planned to modify the original settings to obtain a high-definition image of a thinner layer and a very limited area. If our idea was viable, we'd be able to take a kind of X-ray of the hull through the sediment.

To our delight, the results didn't take long to come. The acoustic images clearly revealed five very narrow breaches, between 5 and 6 metres below the waterline. There may be a sixth, but this has yet to be confirmed by further investigations. The commissions of inquiry were therefore mistaken in asserting that the *Titanic*'s hull suffered a single large tear 100 metres long. Personally, I never believed it. A simple calculation shows such a large opening would have caused the ship to sink in less than ten minutes, instead of two hours and forty minutes. On the other hand, the gouges follow one another in dotted lines for 90 metres, causing water ingress in six out of sixteen watertight compartments. The *Titanic* could have floated for a long time with three compartments flooded. With six compartments affected, she was lost. Even if the cumulative surface area of the breaches were no more than 1 square metre – the equivalent of a closet door – it would be enough to let in 400 tonnes of water per minute. The three bilge pumps operating simultaneously could only evacuate 400 to 450 tonnes per hour. When water flows in sixty times faster than it flows out, the outcome is inevitable. The *Titanic* filled with 34,000 tonnes of water before sinking.

This discovery of a series of breaches tells a rather different story of the shipwreck than was previously accepted. It shows that the liner did not collide with the iceberg, which would have caused a spectacular shock, but only scraped it. Most passengers barely noticed the friction. The *Titanic*'s naval architect, Thomas Andrews, was on board to check on the ship's performance and note any improvements to be made. He was immersed in studying the ship's plans and didn't notice a thing. In the moment, no one was concerned except Captain Edward Smith, who asked Thomas Andrews to come with him to inspect the holds. The two men immediately grasped that the situation was desperate. The first thing to do was avoid causing panic.

In a precise and measured letter addressed as early as May 1912 to the British vice-consul in Cincinnati, a survivor, Mrs Martha Stone[1], wrote: *'I was in bed asleep when the collision occurred. It didn't seem too serious to me, but I did understand that we'd just had an accident. I put on a kimono, fur coat and slippers and went out into the corridor, where I met up with a few people who had come out for the same reason. Someone said we'd hit an iceberg. As a cabin boy was there, I asked him, "What should we do?" He replied, "Go back to bed and go back to sleep, there's not the slightest danger." I saw two of the ship's officers talking. I asked them why the steam was making so much noise. One of them calmly replied, "We've stopped to see what the damage is, which is why we're letting off steam, but there's no danger whatsoever."*[22] [23]'

Such reassuring words would make it difficult to persuade some passengers to leave the comfort of their

---

[22] Martha Evelyn Stone (1850-1924), the wealthy widow of the founder and shareholder of the Bell telephone company, was travelling for pleasure. She occupied the 1st class cabin B28 with her lady-in-waiting, Rose-Amélie Icard.

[23] Extract from the facsimile of the letter, author's personal archive.

cabins to get into the lifeboats and venture out into the freezing night. Sliding down the ship's flank on swinging ropes to the black water 20 metres below must certainly have been a frightening prospect. Some survivors testified that terrified passengers were urged to board the lifeboats but stood paralysed on the lifeboat deck, condemning themselves to certain death. Others survived only because they were literally thrown by force into the lifeboats.

Lifeboat No. 1 left with twelve people on board instead of forty, and lifeboat No. 6 with twenty-four instead of sixty-five. However, the sea was particularly calm, and it would have been possible to slightly overload the boats without undue risk. Only two of the twenty lifeboats were fully loaded. If they had all been filled correctly and on time, the balance sheet could have been reversed: 1,178 people could have been rescued from all the lifeboats, yet only 700 were saved. The *Titanic* had a total of sixteen wooden lifeboats – two lifeboats with forty places (lifeboats 1 and 2) and fourteen lifeboats with sixty-five places (numbers 3 to 16) – and four collapsible lifeboats (A to D). This was 22 per cent more than required by the British Board of Trade regulations, which only obliged a ship like the Titanic to have lifeboats for 962 people. These obsolete 1894 regulations were nonetheless re-validated in 1906. They stipulated that the number of lifeboats be determined by the tonnage of the boat, not the number of people on board. Thus, a ship of over 10,000 tonnes had to be equipped with 16 lifeboats, supplemented by rafts. This ignored the fact that liners routinely reached 40,000 tonnes, with the *Olympic* and *Titanic* exceeding 50,000 tonnes. Fortunately, the *Titanic* was far from full: 1,324 for a maximum capacity of 2,471.

Historians now agree that even if there had been more lifeboats available, they would have been useless, because they wouldn't have been launched in time. Some manoeuvres were catastrophic: collapsible lifeboat A filled up with icy water (many of its passengers froze to death); lifeboat B flipped over. While some officers managed to keep a cool head, the evacuation was chaotic. As a result, lifeboat No. 13 was almost crushed under lifeboat No. 15, which was lowered at the same time.

Mrs Stone's testimony is most explicit:

'. . . the daughter of a lady who occupied the cabin opposite mine came running down the corridor, shouting to her mother that she had to put on her life jacket right away, because we all had to get into the lifeboats.

'No one came to call us or give us orders. We grabbed our life jackets and went up on deck. A steward put them on us, telling us it was just a precaution. [. . .] An officer asked us to go to the upper deck. We did so, and found that they were lowering the lifeboats, but there seemed to be no leader or organization to direct the manoeuvre. I saw no crewmen ready

to take charge of the lifeboats, unlike what I'd seen on other liners during safety drills. There was no sign of panic whatsoever. I saw that there was room in the lifeboat they were lowering and said to my lady-in-waiting, "Let's go now." The sailor in the lifeboat took me by the hand, but the dinghy was rocking so hard that he had to let go to grab a rope and I fell in. [. . .] The lifeboat had been plopped down in such a haphazard way, and swayed so alarmingly that if the sea hadn't been as calm as I'd ever seen it, I'm afraid we would have capsized.[24']

This poor use of the rescue equipment was the result of a whole series of errors. *Titanic*'s sea trials off Liverpool were botched, limited as they were to a dozen hours on 2nd April 1912. The liner obtained her certificate of seaworthiness that evening, before setting sail for Southampton. The second mistake was that many of the crew who embarked in Southampton were completely unfamiliar with the ship, and no safety drills were carried out during the crossing. As a result, the crew encountered serious difficulties in using the new quadrant davits fitted to the *Titanic*. When sailing, the lifeboats were not suspended from the davits[25], but placed on the lifeboat deck. It was up to the sailors to secure the boats and then lower them.

---

[24] *Ibid*, p. 88.

[25] The exception on the *Titanic* were so-called 'rescue' boats No. 1 and No. 2, always positioned to be launched without delay in the event of an emergency (man overboard, for example).

The new system was meant to facilitate and speed up the positioning of the boats along the hull, but lack of training led to the opposite result.

For all these points and many others, Captain Smith carried a heavy responsibility. Not a single safety drill was carried out during the four days it took to sail from Belfast. The one scheduled for Sunday 14th April, just hours before the shipwreck, was cancelled at the last minute because of Sunday mass, an event that could surely have been foreseen . . . Crew members received no instructions about how they would be assigned to lifeboats. During the evacuation, Smith persisted in personally applying the 'women and children first' rule, letting nearly empty lifeboats go rather than allowing men to fill the remaining spaces. Mrs Stone remembered this very well[26]. The results were clear to see. On the starboard side, where operations were led by First Officer William Murdoch under the supervision of First Mate Henry Wilde, the lifeboats left quickly, fully loaded with men, women and children. On the port side, under the supervision of Captain Edward Smith, Second Officer Charles Lightoller was ordered to strictly enforce the rule. The lifeboats were launched late and almost empty.

---

[26] *'The Captain came and stood in front of us, saying, "Women and children first".'* Extract from the facsimile of the letter, author's personal archive.

There were only 500 women and children aboard the *Titanic*. All of them could have been saved, along with many more men. Instead of waiting for the women and children to show up to fill the boats, Lightoller and Smith should have been busy guiding the female passengers, especially those in third class, to the upper deck, which many didn't manage to reach in time. It was easy to get lost in the maze of decks and passageways on the *Titanic*. Rose-Amélie Icard, Mrs Stone's lady-in-waiting, remembers, '*I wanted to go back down to look for Mrs Stone's jewels, worth a fortune. I took the wrong stairs and so came halfway back up. That was lucky, as I would never have got back up again*[27].'

Contrary to popular belief, third-class passengers did have access to the lifeboat decks at the time of the sinking, even though the route was far from simple. This is how the Shuman family (mother and her two children, Eleanor and Harold), who were travelling third class, survived. Eleanor, then sixteen months old, had been told by her mother that a steward had guided her and two other Swedish women from their cabin to lifeboat No.15.

---

[27] Manuscript entitled 'Le naufrage du Titanic' ('The sinking of the Titanic'), written in French on 8th August 1955 by R.-A. Icard who was then aged eighty-three. Facsimile, personal archive of the author.

Another of Captain Smith's errors was maintaining a high speed (22.5 knots, i.e. almost 42 km/hour) even though the Titanic had received a dozen ice warnings from ships in the area during the day. He simply took a slightly more southerly route than planned, believing that this would be enough to avoid the iceberg sector. Ironically, the original route might have allowed the ship to pass unscathed. And then one thing led to another. The boat was going too fast, the iceberg was seen too late, it was a new moon, the night was particularly dark, and the flat calm prevented the white foam that usually signals the base of an iceberg from forming. The officer of the watch, First Officer Murdoch, aggravated the situation by simultaneously ordering the boat to turn to port to avoid the iceberg, and stop the engines before going full astern. The faster the propellers turn forward, the more the ship will veer.

Our natural reflex is to avoid an obstacle, yet some have questioned whether it might have been better to slam straight into the iceberg. Two years later, in March 1914, the liner *Royal Edward* found herself in the *Titanic*'s predicament, facing an iceberg seen at the last moment. Remembering Murdoch's mistake, the captain ordered, 'Go astern with rudder amidships,' to hit it head-on as fast as possible. The forward compartment was crushed, but the ship remained afloat and all 800 passengers were unharmed.

* * *

In the days following the disaster, controversy was rife. Why so few lifeboats? Why didn't the watertight partitions go all the way up to the top of the compartments? Why didn't the radio operators on other boats in the area respond to the *Titanic*'s distress signals? Why didn't the rockets fired from the liner prompt any reaction? Concrete measures were quickly taken. First, the president of the White Star Line ordered the emergency addition of twenty-four lifeboats to the *Olympic*, some of them apparently salvaged from the *Titanic* by the *Carpathia*. The *Olympic* ended up being equipped with sixty-eight lifeboats, compared to the Titanic's twenty. The decision applied to all the company's other liners. So great was the demand for lifeboats that the shipyards couldn't keep up. The following year, the *Olympic* was redesigned in line with the lessons learnt from the shipwreck: full double hull up to G deck, and a new layout for the watertight bulkheads, which were raised to B or D decks.

It was also obvious that the regulations concerning life-boats, in particular those of the British Board of Trade, had not caught up with developments in liner design. By the 1900s, they had become giants of the sea. Most of the great transatlantic liners of the time were over 200 metres long and could accommodate several thousand people.

As for 'wireless telegraphy', it was not yet considered a safety feature, even though as early as 1899, a message sent by wireless telegraphy saved lives after a collision at sea[28]. Wireless telegraphy even played a key role in coordinating rescue operations, particularly when the *Republic* was rammed by the *Florida* in January 1909. An estimated 760 people were saved that day thanks to the radios on the *Republic* and the *Baltic*, two White Star liners equipped with wireless telegraphy. In the aftermath of what could have been a tragedy at sea, Joseph Bruce Ismay, president of the White Star Line, wrote in his notebook:

It's clear that ship safety needs a complete rethink. Today, it is important to build hulls that make vessels unsinkable, and to equip crews with modern means of communication to summon help in the event of difficulties. Lifeboats, on the other hand, are inefficient, outdated and take up unnecessary space on decks, and must be reduced in number. In this way, maximum space can be freed up for commercial use. These principles apply to the two large ships under construction by our company since the contract was signed on 31st July 1908, transatlantic ships that promise to be the most beautiful in the world: the Olympic *and the* Titanic.

---

[28] This message was sent on 30th April 1899 by the *East Goodwin Sands Lightship*, a ship equipped with a Marconi wireless telegraphy set, after it collided with the SS *R.F. Matthews*. The message led to a nearby boat coming to the rescue of the crew.

No comment . . .

The following year, the lifeboat issue led to a conflict between naval architect Alexander Carlisle, who wanted sixty-six lifeboats, and Bruce Ismay, who wanted to limit the number to eighteen. Ismay got his way and Carlisle resigned. Thomas Andrews took over at the shipyard. However, Ismay had been right about 'modern means of communication', but the means were not yet in place. Radio operators were not crew members but employees of the Marconi Company, which supplied the transceivers. Their primary role was to send and receive personal messages, for a fee, for first-class passengers. The presence of on-board radio was first and foremost a business argument. An hourglass (that we found in 1998) was used to time the communications. The two operators alternated during the day and ended their shift at midnight. There was no night shift on any of the ships of that time. Even worse news was that the codes used in emergencies were not standardized. While the 'SOS' distress signal was adopted as early as 1906, the Marconi Company stuck to its own signal, 'CQD' ('Come quick, danger'). The *Titanic* transmitted many CQDs before deciding to send an SOS. However, 'CQD' is easily confused with the CQ code, which merely signals the broadcast of an important message. The CQD code was abandoned after the sinking of the *Titanic*, and a round-the-clock radio watch became mandatory.

* * *

One of the biggest controversies was that a ship seen nearby by many of the survivors was said to have made no move whatsoever to help the victims of the shipwreck. This ship was later identified as the mixed freighter the *Californian*, brought to a halt by an ice field near the *Titanic*. Following an investigation, her captain, Stanley Lord, was accused of failing to go to the aid of the liner, despite the crew firing distress flares. Lord reported that rockets – back then white, not red[29] – were indeed seen at a great distance, but everyone thought they were fireworks on the *Titanic*. On top of which, the *Californian*'s sole radio operator had turned off the radio at midnight, when his shift ended. Lord was accused of dereliction of duty and dismissed by his company, the Leyland Line.

To get to the heart of the matter, we organized our own reconstruction. We placed our two support boats in the respective positions of the *Titanic* and *Californian* and fired white rockets, identical to those used on the liner. We had found a case of them in the debris field – the French manufacturer still existed and had agreed to make a new batch. Our verdict: based on the *Titanic*'s true position, the *Californian* was not within sight of the liner. Although rockets could be seen in the distance, the

---

[29] It has been compulsory for distress flares to be red since 1948.

Californian's crew were not able to deduce their significance. In any case, she was too far away to arrive in time to help, slowed down as she was by an ice field. In light of this, Captain Lord was recently vindicated. Everything points to the mysterious vessel seen by passengers on the *Titanic* being a Norwegian sealer, the *Samson*. Fishing illegally with her lights off, she preferred not to make herself known[30].

After the sinking of the *Titanic*, the British Commission of Inquiry, the Mersey Commission, drew up a long list of recommendations concerning watertight bulkheads, numbers of lifeboats, safety drills, persons in charge in the event of an evacuation, the use of wireless telegraphy, etc. These recommendations became rules after the first SOLAS Convention (Safety of Life at Sea) was drawn up in 1914. Regularly updated under the direction of the International Maritime Organization, the SOLAS Convention still defines the safety and security rules for merchant ships in international waters.

Another consequence of the sinking was the creation of the International Ice Patrol under the authority of the US Coast Guard, to alert ships to the presence and position of icebergs. Today, this is done by aircraft which drop iron

---

[30] See the investigation by American historian David Eno, reported in the following article: Ken Ringle, 'The ship that passed in the night', *The Washington Post*, 30th June 1991.

filings on the icebergs to make them more visible to radar. It's effective, but these floating masses of ice now look like ugly piles of rust.

As well as discoveries about the circumstances of the sinking in 1912, the 1996 expedition was marked by the presence of guests in the *Titanic* zone. The Bass brewery, which had supplied 20,000 bottles of beer for the liner's maiden voyage, had just organized a promotional contest. The top ten prizes were to be awarded during a stay on the expedition's support vessel.

For the ten winners, the trip wasn't exactly a pleasure cruise. Boarding a small ferry to join us, they arrived at night in heavy weather, and had to wait several hours for it to die down before finally being transferred to the support boat. At the prize-giving ceremony, we surprised them with an impromptu exhibition of a dozen bottles of Bass beer from 1912, which I'd fetched from the debris field, especially for them.

This type of stunt helped us finance, at least in part, our expeditions. That same year, 1996, and for the same reason, two ocean liners were on site, carrying a large number of enthusiasts who had come to see us bring parts of the wreck to the surface. The event was covered by live television broadcasts from the site.

Aboard one of the liners were three of the *Titanic*'s last

survivors. Eleanor Shuman, sixteen months old at the time of the tragedy, was saved along with her mother and brother. She was the only one of the three survivors present not to have lost any family members in the disaster. The other two were Edith Haisman and my friend Michel Navratil. One of the most moving moments of the cruise was the instant, captured on video, when Edith and Michel threw flowers into the sea. She threw a red rose, he a multi-coloured bouquet, in memory of their fathers and in homage to the victims. Neither had been back to the scene of the shipwreck since that fatal night in 1912, eighty-four years previously. Back in Montpellier, Michel Navratil sent me this note: 'You've given me the greatest gift I've ever received.'

Before arriving in the *Titanic* zone, Michel Navratil had visited his father's grave in Halifax for the first time in his life. It still lies in the city's Jewish cemetery, the family seeing no reason to move it to the multi-faith Fairview Lawn cemetery, where 121 bodies are buried, some of them unidentified[31]. Seen from the air, the layout of the graves resembles a ship in profile. But that's by accident, not design.

---

[31] The film *Titanic* also left its mark on the cemetery. Fans of James Cameron's feature film are convinced the grave of a certain J. Dawson belongs to the fictional hero played by Leonardo DiCaprio, Jack Dawson. The permanently flower-decked grave is in fact that of Joseph Dawson, a stoker in the boiler room.

Chapter 7

10th August 1998

The Big Piece finally resurfaces

The adventure of the 'Big Piece' unfolded over four years, from its discovery to its removal from the water – four years of effort, disappointment and inventiveness. It all started with a discussion I had in 1993 with George Tulloch. How do we ensure *Titanic* exhibitions remain attractive? Showcasing displays of jewellery, porcelain, crystal and everyday objects such as kitchen utensils, luggage and bottles is all very well and good, but it can become tiresome in the long run. 'Déjà vu', as the Americans say in French. What's more, these artefacts aren't always specific to the *Titanic*, although linked to its history.

We came to the conclusion that we needed to give visitors an idea of the *Titanic*'s size by exhibiting parts of the liner's structure. These pieces are not fragile, so the public can touch them and sit on them, just as a passenger or sailor would have done. From there, we began to locate and bring up some exceptional deck fittings, often weighing in excess of 1 tonne. We developed new techniques for retrieving heavy objects, a far cry from the articulated arm and baskets system used almost exclusively up until then. On 21st June 1993, we hauled up a lifeboat davit (1.6 tonnes); on 23rd July 1994, mooring bollards (2.35 tonnes); and two days later, on 25th July, a roller fairlead (over 1 tonne). But we were hatching a plot – to retrieve a piece of the hull to show how the *Titanic* was built. We just needed to find a part that was actually possible to bring up.

On every dive, the whole team tried to spot a sheet that might be suitable, but the smallest ones likely weighed in at 200 tonnes, and technically that's mission impossible. We finally found an interesting fragment typical of shipbuilding at the time, with its reinforcements and rivets, and several portholes still in place. The piece appeared to be of suitable size and weight, but as it was lying flat, it could have been hiding pieces of deck buried in the sediment. We measured it and I estimated its mass from the thickness of the riveted metal sheets. Around 20 tonnes – that was reasonable enough – and we'd leave it at that for 1994.

* * *

In 1996, we decided that the number one objective of the expedition would be to recover the Big Piece. I ruled out the idea of sending a line down from the surface and attaching it to the part before hauling it up. It's very difficult to lower a line vertically in the middle of the Gulf Stream. It also means tying the surface boat to a 4,000-metre leash in an area with strong currents and frequent bad weather, which could have put it at unnecessary risk. Another solution would have been to moor the rope to floats on the surface, but the likelihood of them drifting and vanishing under the water would be very high. Once again, Pierre Valdy came up with the winning formula: make floats with collapsible fuel tanks – diesel balloons or bags – filled with the boat's fuel. Floatation is achieved by differences in density, diesel being lighter than water.

Each balloon contained 20 tonnes of diesel, giving it a buoyancy of 3.5 tonnes at a depth of 3,800 metres in water at 1°C. Liquids become compressible at these depths and we allowed for this in our calculations. The conclusion was that by using eight of these 'balloons', we obtained a buoyancy of at least 24 tonnes – enough to raise the piece – the exact weight of which was unknown to us. We'd also be facing the challenge of pulling it out of the sediment. As an additional safety measure, we'd invited David Livingstone, chief naval architect at Harland & Wolff, builders of the *Titanic*, to join us on site. In view of the piece to be lifted, and after studying the ship's construction plans, he confirmed that its weight should not exceed 20 tonnes.

Another risk factor was the thermocline, a zone of rapid thermal transition between the very cold water at the bottom and the surface water, which was more temperate at that time of year, thanks to the tropical influence of the Gulf Stream. In the currents zone, the force exerted on the lifting line acts like a sail, making it hard for the floats to surface. The lifting line then has to be attached with a grappling hook, which can be a nightmare to achieve. We'd already tried this out during an earlier operation.

Each bag was weighted down with 3.5 tonnes of old chains, to which shot bags were added to adjust the weight, and the whole thing was dropped into the sea. At the bottom, the effective weight[32] of each bag was between 30 and 40 kilos, as the submersible was not able to lift more than this. The *Nautile* then brought the bags over one by one to connect them to the Big Piece. More chains were passed through the fragment's portholes and then attached to the floats. After eight dives, all the balloons had been moored. All we had to do was release the weights and let the whole thing rise.

---

[32] The weight of the weighted bag minus its buoyancy weight.

The operation took place on Thursday 29th August 1996. I had planned to release the weights one by one to see how the balloons behaved, and then, to be on the safe side, to move the *Nautile*. That day, I was in the submersible. When we started dropping the weights, the first bags began to pull on the sheet metal which, instead of straightening, bent at almost 90 degrees before our very eyes. I immediately thought, we've had it, it's going to break! We alerted the surface crew, and the tension went up a notch! Nevertheless, we decided to give it a go. Early the next morning, when the sea was calm, the ballast weights of the other balloons were released. A few minutes later, we noticed the depth indicated by the navigation beacons supporting the fragment was decreasing. This was a sign that the fragment was lifting off the bottom and beginning its ascent. The balloon's ascent accelerated rapidly, its speed becoming alarming: was the Big Piece still attached? Once the bags were brought to the surface, the tension of the cables showed that the 20 tonnes were indeed still there.

Like all those interested in the *Titanic*, I've read pages and pages of demonstrations where serious naval architects claim that low-quality steel is weakened by low temperatures, making it brittle on impact. But I've seen with my own eyes how, even after more than eighty years in icy water, the steel used for the *Titanic*'s hull remains malleable, being able to bend and then regain its shape despite all the reinforcements behind the piece. At the bottom, I've often noticed sheets torn from the hull and bent 180°, but never broken. Clearly, our experience contradicted tests in the lab.

The blame has often been placed on rivets, some made of steel, others made of iron, therefore containing less carbon. Specimens have been sawn up and examined under the electron microscope for cracks and embedded slag. These analyses have led to some fascinating discoveries about the metallurgy and practices of the period, in particular the quantity and distribution of slag in the metal. This was where the expertise of the smithies resided at the beginning of the 20th century. The riveters at Harland & Wolff used more empirical but effective methods. They knew from the sound of each rivet whether it was of good quality or not. Of course, medium-quality products could escape this type of auditory inspection.

Three million rivets were needed to build the *Titanic*. The riveting technique was simple: if the rivets were made of steel, two rows were fitted; if they were made of iron, three rows were fitted. Whatever the defects that we can now identify in the ship's sheet metal and rivets, they were undoubtedly of the best quality available, and the work was carried out according to the best practices of the time.

It's also worth pointing out the simple fact that no ship is designed to scrape an iceberg at full speed. It would be interesting to see how a current welded-steel liner would perform under the same conditions. Maybe we'd have a few surprises . . . In 2007, a modern cruise ship, the *Explorer*, sank in the Antarctic Ocean after colliding with an iceberg that ripped through its ice-capable hull. There's no comparison between an iceberg and ice cubes from the freezer. An iceberg is made of extremely hard, compressed freshwater fossil ice that has fallen into the sea due to the advance of glaciers. The one struck by the *Titanic* weighed an estimated 1.5 million tonnes. When a 50,000-tonne ocean liner travelling at 22.5 knots[33] scrapes a reinforced concrete quay – or an iceberg – it's the same thing; it's inevitable that the rivets will break or the sheet metal will give way, no matter how good they are. No ship is unsinkable, even today; they're just less vulnerable.

As the Big Piece neared the surface, our worries began. The piece was to be hoisted aboard the *Jim Kilabuk*, an ocean-going supply vessel with a long, unobstructed quarterdeck that was flush with the water. The boat was supposed to be specially equipped with a construction crane capable of lifting 20 tonnes. Before the expedition,

---

[33] More than 40 kilometres per hour.

I discussed the matter with the Canadian Coast Guard, who agreed to the use of this hoisting system on board. But when the *Jim Kilabuk* arrived the day before the operation, well behind schedule, it was not fitted with a construction crane, but a simple fixed gantry crane. The crew was optimistic but I was far less so, especially as there was no cable on the gantry pulley. In action, the pulley proved totally useless – it was inaccessible when the boat was at sea, making it impossible to run cables through it. We had to make do with simply winching items onto the cable guide rolls located at the end of the quarterdeck. Then, we encountered yet another problem – the roller jammed. The tension was such that a first balloon cable broke, then a second. The piece plunged back down, still suspended by the other balloons. We had to urgently stop operations or risk losing it.

That day, my son Julien was on a Zodiac near the *Jim Kilabuk*. The Ifremer divers present were willing and able but were not authorized to dive. So he did an air dive[1] to check at what depth the Big Piece was suspended. When he came back up, he told me he'd been down to –70 metres but couldn't see it, even using a powerful lamp (I actually always thought he'd dived a little deeper). I momentarily considered diving down to try to attach a hook, but going down alone – probably to almost –100 metres – and having to make the physical effort to moor the fragment was too risky, especially as night was falling. So we took the only sensible decision: to head for the shallows of the Grand Banks of Newfoundland and drop the Big Piece at a depth of 50 to 60 metres – less if possible – where it would be easy to dive and prepare for a fresh ascent the following day.

The distance to be covered was no more than 80 nautical miles[34] [35], or around twenty hours' sailing at 4 knots cruising speed. But our run of bad luck was far from over. At around 3am, we were caught in Tropical Storm Edouard, with 1.5-metre-high waves sweeping across the quarter-deck. It was impossible to lash the Big Piece down more securely, and its floats slapped against the *Jim Kilabuk*'s hull. The repeated impacts broke the hoisting lines, and the piece sank after drifting 10 miles, or almost 20 kilo-metres, coming to rest like a sword in the sediment at a depth of 3,200 metres. Fortunately, there was still a navi-gation beacon on one of the last floats, allowing us to track the Big Piece's descent and position from the *Nadir* throughout the night. But the disappointment was

---

[34] When you dive with compressed air cylinders, nitrogen starts to be toxic at ‾40 metres and oxygen becomes dangerous at ‾85 metres. Diving with air to ‾70 metres and more is exclu-sively for highly-trained divers. It has been off-limits for professional divers since the late 1970s.

[35] 150 kilometres.

immense – enough to make a grown man cry – and many of us had tears in our eyes.

The next day, I dived in the *Nautile* (George Tulloch didn't have the heart to dive that day) to assess the situation. The Big Piece was in the position indicated by the beacon, and the mooring chains passed through the portholes had not moved, which would facilitate ascent operations. We collected the floats and beacons, which are expensive items, as we weren't sure of returning. We left behind us a plaque with an optimistic message: I WILL COME BACK, signed George Tulloch.

In financial terms, the failed mission was very bad news. The aborted ascent of the Big Piece cost RMS Titanic, Inc. several million dollars. A large part of this was financed by the sale of 2,000 tickets, costing between $1,500 and $5,000, to watch the event from two luxury cruise liners[36], the *Royal Majesty* and *Island Breeze*. On both boats, each cabin was equipped with closed-circuit television so that passengers could follow the expedition's progress live.

We'd just spent ten dives preparing for this ascent, and everything had fallen apart at the last minute . . . Like a good sport, George Tulloch told the media, 'The ocean gives no quarter. This attempt failed because we neglected

---

[36] See previous chapter.

to coordinate the winching and hoisting technology of the 21st century with that of the 19th. We won't make the same mistake twice.' Behind the scenes, George was less philosophical. In fact, he was furious. Like the rest of us, he knew responsibility for the failure lay solely with the owner of the *Jim Kilabuk*, who breached his contract by sending a poorly maintained and under-equipped boat at the last minute, with an incompetent but self-assured operations manager.

Do I need to spell it out? We refused to pay his bill!

Two years later, we returned with more powerful and better adapted equipment, designed by Pierre Valdy. This time, the *Abeille Supporter*, with its large mobile gantry crane, would hoist the Big Piece. The first step was to place a field of beacons around the site of the fragment. We found it on the *Nautile*'s first dive. After a series of dives to attach the balloons to a single heavy gauge hoisting line and methodically prepare for the ascent, the big day finally arrived.

On Monday 10th August 1998, the weather was bright and sunny, so we decided to release the ballast weights early in the morning and make the most of the long day's work ahead of us in natural light. The weights were dropped, I scanned the sea from the *Abeille Supporter* and, twenty minutes later, the balloons appeared on the surface.

It all happened so fast, I was convinced there was nothing at the end of the line. But there was! It was an image I'll never forget. It was as if we were seeing the *Titanic* starting to surface as the first centimetres of sheet metal came into view.

Once it was hoisted onto the *Abeille*'s quarterdeck, I was finally able to examine the Big Piece from every angle. This steel rectangle measured around 7 metres long by 4 metres wide, and a 3-metre-long triangle extended downwards – a vestige of the lower level. It retained a perpendicular beam almost 1.5 metres long, which still supported a fragment of the deck. This spar, planted in the sediment, was invisible when we chose to bring up the Big Piece, and could have made life difficult, but the operation went smoothly.

Once the Big Piece had been brought up, the most important work began: treating it to halt its deterioration before exhibiting it. Carried by the *Abeille Supporter* to Boston harbour, the piece was entrusted to EverGreene Architectural Arts, the company charged with restoring it. Treatment consisted of neutralizing chlorides in an alkaline bath; mechanical cleaning; anti-corrosion treatment with tannic acid; and application of microcrystalline for protection and finishing. This treatment took place over two years, from 1998 to 2000. In 2002, however, corrosion once again raised its ugly head. After a year of analysis and study of the first restoration process, a new treatment was deemed necessary. A different method was used from 2003 to 2004. The remaining rusticle crust was removed using an ultra-high-pressure cleaner (up to 3,000 bar). The exposed sheet metal was left in the open air to rust naturally for two days, before being brushed with a phosphoric and tannic acid-based rust converter. Finally, a new layer of microcrystalline wax containing corrosion inhibitors was applied under heat. The Big Piece now seemed to be stabilized[37].

For ease of handling, the Big Piece was cut into two pieces, the lower triangle having already been detached from the main piece. These two pieces were featured in the two permanent exhibitions of RMS Titanic's artefacts: the larger in Las Vegas (Nevada); the smaller, in Orlando (Florida).

The Big Piece is without doubt the most fascinating artefact to be recovered from the *Titanic*. Visitors can see a piece of the liner's hull exactly as it looked to the curious onlookers and passengers who crowded the docks before its departure. It has four portholes, all fairly well preserved – two large cabin portholes and two small

---

[37] See Joseph Sembrat, Patricia Miller, Justine Posluszny Bello, 'Conservation of the RMS Titanic 'Big Piece': A Case Study and Critical Evaluation,' *APT Bulletin: The Journal of Preservation Technology*, vol. 43, No. 4, 2012.

bathroom portholes. It belonged to first-class cabins 79 and 81 on C deck, unoccupied during the 1912 crossing.

Curiously, when I spotted the Big Piece on the sediment in 1994, the four portholes were already open, which made it much easier for us to pass the hoisting chains through. It's quite astonishing to see how many of the *Titanic*'s portholes were wide open. This, despite it being very cold on the day of the sinking, and the fact that portholes are normally kept closed at sea. Were they opened when the ship began to sink? And why? This is one of the questions that remains unanswered.

Every time I see the Big Piece, I remember the shock we felt when we found it planted vertically in the sediment. As we passed astern with the submersible, seeing the portholes on the cabin side and the small section of deck still in place, for a second, I had the weird sensation of entering the interior of the *Titanic* for the first time. This is undoubtedly the same feeling visitors have when browsing through the *Titanic* exhibits: reliving not just the sinking, but also something of the atmosphere of the great transatlantic liners of the early 20th century. And in the process, discovering the exploration and conservation techniques that allow such impressive relics to be saved today.

Chapter 8

2010–2019

What future for the Titanic?

At the end of 1999, on Thanksgiving Day, my relationship with RMS Titanic was interrupted by a
bolt from the blue. In the wake of a takeover, George
Tulloch was ousted from the company and replaced by
a triumvirate drawn mainly from the entertainment
world. The company had big plans. In 2000, it would
mount 'the most ambitious deep-sea exploration and
recovery mission ever undertaken', according to Arnie
Geller, its new president. His primary objective, which
he loudly proclaimed in the media, was to recover the
$300 million worth of diamonds that he claimed were
languishing in the cargo hold. His attitude was in total
contradiction with the rules Tulloch and I had set
ourselves, so I decided to distance myself. I was to be

proved right. Expedition 2000, hastily put together at a high price (we're talking five million dollars), monopolized some fifty people, including a dozen scientists, an oceanographic vessel and two exploration submersibles. One part of the team was American, the other Russian, and the language barrier complicated operations.

It didn't come close to its objectives, but the 2000 expedition nonetheless brought back 800 artefacts: crockery, shoes, documents, tools, a fine collection of postcards, and some very rare perfume samples, some intact, others broken. The diamonds, if they ever existed, remained at the bottom.

During those years, I managed Aqua+, a subsidiary of Canal+, whose mission was to make underwater films. Aqua+ owns a floating research and studio vessel, the *Ocean Voyager*, a helicopter and two single-seater Deep Rover submersibles. I dived on two wrecks, the *Royal Captain*, a 15th-century English ship, and the Japanese battleship *Yamato* in the China Sea.

The year 2004 opened with sad news. George Tulloch passed away on 21st January, struck down by cancer at the age of fifty-nine. This passionate *Titanic* enthusiast had an unquenchable thirst for learning. He who had never sailed before (he owned a small pleasure boat which he very seldom used) took to the sea and deep-sea diving like a fish to water. The president of the Titanic International Society, Michael Findlay, said of him: 'With the Big Piece, he literally brought the *Titanic* back to port. He did more than anyone else to preserve the memory of the *Titanic*.' I liked the entrepreneur, I liked the man, and we were very close from 1987 until his passing.

2004 also saw a fresh expedition to the wreck, organized by RMS Titanic, Inc., using a Remora 6000 ROV (remotely operated vehicle) supplied by Phoenix International, a company specializing in underwater operations. Seventy-eight artefacts were recovered from the debris field during the ROV's ten dives between 28th August and 27th September.

For me, 2007 marked the renewal of good relations with the new RMS Titanic, Inc. management team (the 2000 team had dissolved) and saw me directing a campaign on the wreck of the *Carpathia*. This was the ship that saved all the *Titanic*'s survivors and was later torpedoed by a German submarine in July 1918 south of Ireland. This campaign used the resources of the French company, Comex.

The adventure resumed on a whole new scale in 2010. I had been commissioned by the Bureau of Enquiry and Analysis for Civil Aviation (BEA)[1] to lead one phase of the search for Airbus flight AF447 [38] [39], which had disappeared off the coast of Brazil. Shortly after this operation, RMS Titanic, Inc. entrusted me with the preparation and execution of a new campaign on the wreck of the *Titanic*, which I hadn't seen for twelve years. Initially, we planned to use the *Nautile*, but it was not available, so we opted for Russian Mir submersibles. The contract was as good as signed when the person in charge, my friend Anatoly Sagalevitch, contacted me in distress. He told me there was a problem with the surface vessel, the *Akademik Mstislav Keldysh*. It was on a long-term charter from Fugro, a Dutch geotechnical company that would later search for flight MH 370[1]. We quickly decided to use the same equipment as for AF447 (I had thought of this during the search for the plane), replacing the US Navy's resources with those of Phoenix International, while keeping the same operators. AUV drones [40] [41], owned by the Waitt Institute

---

[38] French organization under the authority of the Ministry of Transport.

[39] Air France flight 447 (Rio-Paris), an Airbus A330-200, crashed into the Atlantic on 1st June 2009. The wreckage of the plane was found on 3 April 2011 at a depth of 3,900 metres.

[40] Malaysia Airlines flight MH 370, a Boeing 777, went missing on 8th March 2014. The wreckage of the plane has never been found.

[41] *Autonomous Underwater Vehicle.*

of Discovery[42], were operated by a team from Woods Hole Oceanographic Institution. Things happened so fast that the drones returning from Brazil went straight to Newfoundland, where they were returned to operational condition on board the expedition ship.

The expedition involved RMS Titanic, Inc., Woods Hole, the Waitt Institute, Phoenix International, Inc. and a guest archaeologist from the US National Oceanic and Atmospheric Administration. In terms of technical resources, we had the oceanographic vessel *Jean Charcot* – a former Ifremer vessel – two Remus 6000 AUVs and the Remora 6000 ROV. Our number one objective was to map the entire area of the *Titanic* wreck with high-definition 2D and 3D photographic and video images, and acoustic imagery using multibeam sonar and side-scan sonar.

Acoustic maps were taken at an altitude of 80 to 120 metres above the seabed, while photographs were taken at an altitude of only 7 metres, each image covering an area of 7 by 7 metres. This gave us 130,000 photos of the debris field. The entire wreck was filmed in 4K – quality video from the side and from above. Once assembled, the photo-mosaics made from the videos showed panoramic views that are impossible to obtain on site, given the short

---

[42] The Foundation for the Preservation of the Oceans created by businessman Ted Waitt, founder of Gateway computers.

range of the submarine's searchlights. All these images can be zoomed in on to study specific features, such as rusticles and their orientation in current zones.

2D and 3D cameras were built on board before departure. We had the *Jean Charcot* at our disposal for a limited period. It was between two operations for a company that was chartering it and which shared the cost of hiring it.

At the end of this data-rich campaign, we finally had a detailed overview of the site. For years, this general view had been lacking, making it difficult to know precisely which areas had been explored and which had been unintentionally left out, despite all our efforts to locate them. As the AUVs worked very quickly, we decided to extend the search well beyond the known limits of the debris field to check that no important items had been missed, but there was nothing of interest to be seen.

This ultra-precise mapping makes it possible to visualize the distribution of objects on the seabed and measure their orientation and the exact distance between them, providing invaluable information. We took this opportunity to report all known positions of artefacts recovered in previous years, paving the way for an archaeological report on the wreck.

The full sonar map was georeferenced – each time you click on a point, its longitude and latitude are displayed.

This meant we could accurately measure distances between the front and rear of the wreck, and between objects and debris. It's slightly less precise than GPS, but in practice, on the seabed, accuracy equal to the viewing distance, i.e. 5 to 10 metres, is more than sufficient. And 2D maps are ultimately more accurate than 3D maps, which amplify relief to the point of distorting reality and coming to false conclusions, as happened with the compression of the boilers.

In future, these detailed maps of the area will enable the surface boat to guide submarines with great precision. It is impossible for a submarine that's submerged to pick up a GPS signal directly, let alone at a depth of 3,800 metres.

This imagery also provided a wealth of information on the deterioration of the wreck, the subject of much discussion since the discovery of the famous rusticles. This neologism, coined by Bob Ballard from the English words *rust* and *icicle*, refers to the rust concretions, similar in appearance to underwater stalactites, that encrust the *Titanic*'s hull. The French Canadians call them 'rouillons', but I prefer to say 'stalactites d'oxydes de fer' (iron oxide stalactites), even if that's a bit long. Their study has revealed that this is not a mineral phenomenon, but an animal one.

Rusticles are living entities, made up of millions of nutrient-carrying channels, colonized by communities of anaerobic halophilic fungi and bacteria (i.e. salt-loving bacteria that thrive in the absence of oxygen) that feed on iron oxides. These formations are not exclusive to the *Titanic*. They can be found on other deep-sea wrecks, including the German battleship *Bismarck*, sunk in 4,800 metres of water 400 nautical miles off Brest. The rusticles on the *Titanic* replace themselves rapidly, probably in response to the currents, as has been observed since the discovery of the wreck in 1985. Curiously, those on the Bismarck, on the other side of the Atlantic, show the same evolution over the same period.

Rusticles grow at a rate of around 1 centimetre a year. Their life cycle is estimated at between five and ten years, after which they break off in patches under their own weight and the action of the currents. They leave the sheet metal nice and clean until the next 'growth spurt'. They are extremely fragile, disintegrating into fine dust at the lightest touch.

Biologists are fascinated by such formations and have tried to acclimatize them to the surface in their laboratories. The first person to successfully reproduce them in an aquarium was Canadian marine biologist Roy Cullimore. He gave a public demonstration of his work at an exhibition of *Titanic* artefacts in Hamburg in 1998, where visitors could observe colonies of bacteria-building rusticles. A laboratory in Grenoble is doing similar work. This piece of research identified twenty-seven different species of bacteria on the *Titanic*, including one specific to the wreck, called *Halomonas titanicae*.

Since 1994, bacteria traps have been placed on and around the wreck. The first trap, designed by Roy Cullimore, consisted of photo films installed on the deck of the wreck. Roy Cullimore found that the silver metal in the film had been 'eaten away' after a few days, revealing the presence of bacteria. As the expeditions progressed, the traps were perfected. In 1998, four test structures made of different steels were placed on the wreck. Unfortunately, these traps were deliberately dumped outside the wreck by an expedition which included scientists with scant respect for other people's research . . . These 'bacteria traps' have been reinstalled, some replaced by new ones. They will be collected as soon as possible.

In 2019, a diving campaign on the *Titanic* would take place in a very different context from previous expeditions. It was added on to The Five Deeps expedition, led by American explorer Victor Vescovo and his company Caladan Oceanic in 2018. Victor Vescovo had his own equipment, the *Pressure Drop* support boat and the two-seater *Limiting Factor* submersible[43], capable of

---

[43] Built by Florida-based Triton Submarines.

descending to 11,000 metres. His ambitious goal was to dive into the five deepest trenches in the five oceans: Puerto Rico Trench in the Atlantic (8,376 metres), South Sandwich Trench in the Southern Ocean (7,434 metres), Java Trench in the Indian Ocean (7,192 metres), Mariana Trench in the Pacific (10,934 metres) and Molloy Deep in the Arctic (5,551 metres). Incidentally, a scientific lander, stuck at a depth of 10,923 metres in the Mariana Trench, was brought up in the deepest recovery operation ever carried out by a crewed submersible.

Other dives were scheduled in the shallower Tonga trench, as well as on the Titanic – a goal of one of The Five Deeps expedition's first descents. Arriving on site in the summer of 2018, we stayed for a week, but terrible weather, caused by a low-pressure system trapped in the zone, made diving impossible. It was when the expedition returned in August 2019 that we went back to the Titanic. This time, the mild weather allowed us to make five dives in a week. We planned to bring up the 2010 bacteria traps, which were of different shapes and materials, so we could identify the surfaces and metals most favourable to the development of rusticles. We also wanted to replace them with new traps. But a problem with our manipulator arm prevented us from carrying out all the operations.

* * *

This short expedition gave me a chance to observe the deterioration of the wreck, which I hadn't seen for nearly ten years. The forward section had taken a beating during the sinking. The bow was planted like an arrow and extended for some 40 metres to the foot of the gangway, and the 80 metres of hull between the gangway and the second funnel were cantilevered. This section had collapsed onto the bottom at an 11-degree angle, completely twisting the hull. In 1987, a gaping crack in the hull was visible on the starboard side at the point where it had folded. From 1996 onwards, the port side began to blister and, in 1998, to crack. The fracture was now highly visible, about twice the diameter of a porthole, so 80 centimetres to 1 metre wide, and spreading downwards. The sheet metal was so brittle that a porthole had broken off en bloc and was about to fall. Over time, the section of the loading deck that was damaged when it hit the bottom would subside. The rest of the front section offered no easy way to get inside, especially around the grand staircase. In a century or more, the front will look just like the back. The decks will collapse and pile up like the puff pastry ad cream in a *millefeuille*; the hull walls will open up and the whole area will be flattened, as is already the case with the stern, some of whose superstructure is now below the level of the main deck.

This progressive opening up of the 'V' of the hull and the general flattening of the decks is consistent with the normal degradation process of wrecks, with or without rusticles. But to say that the *Titanic* will vanish within ten or twenty years . . . That seems a gross exaggeration, especially as this claim has been repeated for over thirty-five years. The last I heard, the wreck is still there. It's estimated that the bacteria 'eat' between 250 and 300 kilos of metal per day, which seems enormous. But given the 50,000 tonnes of steel on the liner, total disappearance will not occur for four or five centuries, which leaves a decent margin. Not to mention that parts like the boilers and engines are colossal.

The finger has also been pointed at underwater tourism, which some say is damaging the wreck and polluting the site. Anyone would think the *Titanic* was being visited by hordes of unscrupulous divers, all taking bits and pieces as souvenirs and leaving detritus in their wake. This does not reflect reality. Descending by submersible to a depth of almost 4,000 metres is not within the reach of just any diving group. Collecting artefacts requires sophisticated tools, and there has been an average of one expedition every two years since 1985. Some of these were carried out solely by robots or observation submersibles with no possibility of bringing up artefacts. As for the notion of damage being caused by the weight of submersibles landing on superstructures and crushing them, this is the stuff of fantasy, or rather ignorance of the most elementary laws of physics. Out of the water, the *Nautile* weighs 18 tonnes. Once submerged, its effective weight is zero due to Archimedean buoyancy. A submarine landing on the *Titanic* would, at worst, exert a pressure of 4 or 5 kilos over an area of several square metres, which is virtually nothing. Commander Smith's famous bathtub is also said to have disappeared. In reality, it is still there, hidden under a pile of recent debris.

Faced with the reality of the wreck's inevitable disappearance, people have come up with the most colourful solutions. Robert Ballard even suggested repainting the ship! This was most likely a joke, but many took it at face value and went crazy on *Titanic* forums and in the media. Considering the challenges of stripping and treating the Big Piece, which measures less than 30 square metres, the very idea of brushing and then painting over a hectare of sheet metal at a depth of 4,000 metres is simply crazy. And don't forget, the interior of the wreck would also have to be painted . . . And let's not even get started on the pollution.

The fitting of anodes is also envisaged to slow down corrosion by limiting electrolysis. The process, currently being tested in Europe and the United States, was designed primarily to slow down the degradation of wrecks containing dangerous cargoes – effectively chemical 'time bombs'. Even if the tests were conclusive, this would not save the *Titanic*, as several factors contribute to the slow destruction of the wreck: bacteria, electrolysis, oxidation, currents . . . It would be impossible to stop all those processes. Wrecks are destined to gradually disappear. The *Titanic* vanished from the surface long ago. It will also disappear from the bottom, full stop.

Repainting the *Titanic* is just one of the many far-fetched projects the wreck has inspired since its sinking. Ideas for refloating the wreck have included the use of giant magnets connected by cables to winches or floats; filling the wreck with a lighter-than-water hydrocarbon-based jelly, or with millions of ping-pong balls; and enclosing the wreck in a giant iceberg that would rise to the surface with the liner. I confess to having had my own share of dreams, even pondering, along with one of the directors of RMS Titanic, Inc. and in a strictly private capacity, the possibility of bringing up the bow of the *Titanic* in one piece. At that time, there was a ship in the United States, the *Glomar Explorer*, that had been used for an ultra-confidential mission in the 1970s to recover a Soviet submarine loaded with nuclear torpedoes when it was lost at a depth of 4,000 metres. This boat resembled an offshore drilling vessel, but three times larger, and its drill pipe was fitted with a gigantic clamp capable of pulling the submarine up in one piece.

This extraordinary operation was only half successful, as partial failure of one of the clamp jaws caused the hull to break in two. But the stern, which contained all the Soviet navy's confidential codes, was brought to the surface with the utmost discretion. After a long period of disuse, the boat was put up for sale at scrap metal prices. A moment of temptation . . . On reflection, though, the cost and technical complexity of refitting it were inordinately high. In any case, the bow section of the *Titanic* had been too badly damaged by the sinking to be brought up without breaking into at least two pieces. Technical problems aside, what use is a half-*Titanic*?

And yet, the wreck's future isn't completely doomed. In the course of our expeditions, we'd seen something more positive than the prospect of its collapse. And that's the marine life it hosts. I often refer to shipwrecks as 'public housing for fish', and the *Titanic*, which landed on an underwater plain as bare as a football pitch, quickly became a refuge for marine animals despite the depth. Twenty-four different species of fish, crustaceans, sea anemones, ascidians and corals colonize the wreck. We encountered squat lobsters and crabs – all white, since they live in permanent darkness – many varieties of starfish, especially brittle stars, and strange creatures such as chimaeras, those tail-less, inquisitive fish that circle the submarine. Viperfish, too, with their oversized jaws and teeth. Most surprising is the presence of a fish from inter-tropical zones, the tripod fish, perhaps carried by the Gulf Stream. In addition to its pectoral fins, it has three long rigid fins, two pelvic and one caudal, which enable it to assume a 'standing' position 20 to 25 centimetres above the bottom. Hence the name tripod fish. Once in position, it waits motionless for food to fall. All these animals are small, no bigger than 25 centimetres, except for the chimaeras. Contrary to popular belief, no shark can be found at such depths. As far as I know, they don't venture below 2,000 metres.

Some specimens of the *Titanic*'s fauna have never been seen elsewhere. One example is a previously unknown sea cucumber with a lavender-coloured body punctuated by lateral phosphorescent rings. Since the discovery of the wreck in 1985, there have been significant changes in the ecosystem. Eleven years later, the population of brittle stars and sea cucumbers had increased by 75 per cent, while crinoids and ascidians were colonizing the seabed. Magnificent corals have developed on the stern, and I've been watching them grow for the last thirty-four years. Red shrimp can be found here, and numerous 'nests' of black pebbles have been built by an as-yet-unknown creature.

Analysis of the seemingly sterile sediment reveals the presence of around a hundred different animal species.

Should this explosion of life around the wreck be seen as a consequence of overfishing or pollution, with the discharge of fish waste into surface waters increasing nutrients reaching the bottom? Or is it due to a change in the surface currents bringing more plankton into the area? These are all hypotheses that require confirmation, especially as no one really knows what's going on elsewhere, simply because they haven't been there. But it's clear that the wreck of the *Titanic* today represents an oasis in a vast desert.

Epilogue

Unlocking the Titanic's last secrets

The *Titanic* 2021 expedition, dedicated to observing the wreck, used a new type of submersible, the *Titan*, with a carbon fibre hull. It's the only submarine in the world capable of carrying five people – a pilot, a co-pilot and three passengers – to a depth of 4,000 metres. It belongs to OceanGate Expeditions, which runs deep-sea discovery missions open to non-professional underwater explorers. I was invited by the company to join the expedition as a wreck specialist, as were other *Titanic* experts, divers, archaeologists, scientists and historians. This expedition will not be the last – more dives are scheduled from 2022 onwards.

Between late June and early August 2021, the *Titan* carried out five missions, enabling some twenty enthusiasts to visit the wreck. Some were very wealthy, others had made big financial sacrifices, but all were making a dream come true – seeing the *Titanic* with their own eyes. Unlike submersibles designed for scientific use like the *Nautile*, the *Titan* features a single large porthole measuring 53 centimetres in diameter, through which the three passengers can see the underwater landscape at the same time. Unusually, there was a total absence of current, so we could observe the wreck from unfamiliar angles, float over the entire longitudinal axis of the bow section without drifting, and examine the break in its rear section. The most corroded metal sheets had now fallen away, along with a number of portholes, providing unprecedented glimpses of the wreck's interior, in particular the first-class library. All the books are, of course, gone. As I'm lucky enough to know the wreckage and debris field inside out, I was able to comment on what was unfolding before our eyes as we went along. I found myself playing the role of *Titanic*'s underwater guide.

In thirty-four years of expeditions, examining the *Titanic* from every angle, photographing, filming, sonar scanning and exploring the interior of the wreck with robots capable of squeezing through the tightest spaces, we've learnt a lot about the liner and her final moments. But above all, we know that we still have a lot to learn about the wreck of the *Titanic*, as we do about all the world's oceans and seas.

The next few years could be used, on the one hand, to bring up emblematic parts of the ship that are key to its history, and on the other, to finally solve the enigma of its sinking by taking a closer look at the damage caused by the iceberg. In 1996, a survey of the starboard bow with a sediment sounder revealed five, possibly six, small breaches that had opened up on the hull. Although we did some genuine scientific work, this study could be repeated using a more powerful sediment-penetrating sounder, giving colour and 3D images to complete the results. I'd also like to try and see the damage with my own eyes rather than on a screen. I'd like to be able to slide a metal blade into the gaps to see if it cuts through the hull, and check if the sheet metal has torn or if the rivets have popped out. Logically speaking, it has to be the rivets. But after 110 years of more or less serious but contradictory hypotheses, it's time to find out for sure.

Using this 3D sounder, we could explore the stern of the wreck and, more precisely, the part of the stern buried in the sediment. This would make it possible to visualize the central propeller and determine whether it has three or four blades. The Harland & Wolff shipyard register shows four, but the entry is crossed out and overwritten with the hand-written words 'three blades'. Yet another mystery to unravel.

The whole rear section, although badly damaged, deserves to be examined in greater detail now that the most dangerous sheet metal has been removed. Even though detailed plans are available, it would still be exciting to see the Parsons steam turbine up close if it became accessible. This monster generated 16,000 horsepower and used steam at the end of the cycle[44] to drive the central propeller (in forward gear only) and gain a few extra knots without increasing fuel consumption. In places, the anti-roll keels seem to be visible, but this may be an optical illusion created by a crack in the hull, which is beginning to sag. Anyway, it's well worth a look.

Among the key historical items still to be brought up are the radios. They would already have been recovered had the 2021 expedition not been cancelled due to Covid. These Marconi devices were legendary in the history of radio. They played a key role in the outcome of the shipwreck: the 700 or so survivors owed their lives to them. Without the prompt arrival of the *Carpathia*, alerted by *Titanic* radio-telegrapher Jack Phillips, many would have frozen to death in the lifeboats. Equipment that has saved hundreds of lives deserves to be rescued. And rescue is all the more urgent as these devices are located in an easily

---

[44] On the *Titanic*, steam produced by the boilers could be used four times: in the high-pressure, medium-pressure and low-pressure cylinders of the reciprocating engines, and finally in the turbine. After the exhaust steam condensed, the water was returned to the boilers for a new cycle.

accessible part of the wreckage, which is most likely to collapse in the short term.

We could also bring up some major mechanical fragments scattered across the debris field: a reciprocating machine cylinder, one of the auxiliary boilers, crank arms. We've already recovered a shaft bearing with a diameter of 1.2 metres. The auxiliary boilers weigh around 60 tonnes, and the first two machine cylinders weigh around 70 tonnes with their connecting rods. A few years ago, tackling such monumental pieces would have been unthinkable. But today, with new lifting equipment, we know it's possible. The dream would be to bring up a propeller, as was done on the *Lusitania*. But the *Titanic*'s two lateral propellers, the only ones partly visible, weigh 38 tonnes for a diameter of 7.2 metres and, crucially, remain attached to their drive shafts, which weigh 118 tonnes for a diameter of 67 centimetres. As the bolts have been welded by corrosion (the propeller hubs are steel; only the blades are bronze), the only realistic solution would be to cut the shafts. A huge undertaking, even if they were hollow . . .

The front holds also deserve to be explored in greater detail. William Carter's famous Renault Type CB 1912 is nowhere to be found. This passenger survived with his entire family; only his car was lost. The bodywork's thin sheet metal and wooden frame, damaged by the impact and corrosion, have long since turned to dust. But we can still hope to spot its chassis and engine. Other holds are filled with the trunks of first-class passengers, trunks that were thrown around during the shipwreck and which have, over time, opened up and let their contents, mainly clothes, escape.

Another plan was to re-examine the door to the strong-room between holds two and three, the object of all our fantasies . . . The *Titanic*'s archives hold a bill of lading, i.e. a list of what was brought aboard. But, as is often the case on ships, not everything was declared. Either the list is incomplete, or the real list remains confidential.

The RMS *Titanic* carried 3,364 bags of mail. Five postal workers sorted the seven million letters during the crossing. After the collision with the iceberg, they scrambled to save the 200 bags of registered mail that might have contained valuables. Aided by a steward who survived, they worked tirelessly and all perished in the wreck. As mail remains the property of the postal services until it is delivered to the addressee, the letters and parcels aboard the *Titanic* remained the property of the British Royal Mail, before being transferred to the United States Post Office (US Mail)[45]. In recent years, however, Royal Mail has become

---

[45] In the US, even private mailboxes, once they've been installed, become the property of the federal government.

a conglomerate of private listed companies. So who owns the mail still aboard the *Titanic*? For over a century, wild rumours have been circulating about registered mail containing diamonds, securities, etc.

One of the directors of RMS Titanic, Inc. always told me that he had documents from certain families proving that there were very valuable objects on board, but I have never seen these documents. The one thing that's almost certainly down there is an illuminated manuscript on parchment of Omar Khayyam's *Rubaiyat*, purchased by an American bibliophile who had an exceptional binding made for it in London. The work, protected by an oak case, was covered in Morocco leather embroidered with three peacocks doing cartwheels in gold thread, and a Persian oud in wood and ivory marquetry. The binding was inlaid with at least a thousand precious stones (emeralds, rubies, amethysts and topazes), each set in a gold mount. Everything points to the book having been taken on board, but no one knows whether it was kept safe in the vault, in the captain's safe or elsewhere. If his description is accurate, it's possible that at least part of the binding has withstood immersion in seawater.

Another magical place that we still haven't managed to see, except via robot, is the *Titanic*'s famous grand staircase. The wooden staircase itself was blown out during the descent to the seabed. Its location is too narrow to enter with a crewed submarine. Unless, that is, the new generation of acrylic glass bubble submarines, such as the *Triton*, lives up to its promise and finally makes it possible to sneak into cramped but not dangerous parts of shipwrecks, whether the *Titanic* or other vessels.

A phenomenal number of artefacts remain to be discovered in the debris field – between 15,000 and 20,000, according to estimates. Some are still unknown to us, others have been spotted once by chance. When moving an ascent balloon towards the Big Piece in preparation for bringing it up[46], I spotted a complete porcelain dinner service of at least 150 pieces. It probably came from a crate shipped from Great Britain to the United States. The crate has disappeared, but the service remains at the bottom, waiting to be retrieved.

The more artefacts and information we gather, the more the *Titanic* appeals to a wide range of audiences. Some are captivated by the technical aspects, as this was a new type of boat and shipbuilding was at a pivotal time of experimentation with compartmentalization, watertight bulkheads and double bottoms. Others are researching the phenomenon of emigration to America through the role played by the great ocean liners. Still others are fascinated

---

[46] See chapter 7, p. 103.

by the stars of the era on board the ship, or by the stories about the passengers, the survivors and their families. There's something to interest everyone in the story of the boat and the shipwreck. This is not the case with many other catastrophes, where the tragedy itself puts an end to the story and it fades into the mists of time.

If I have one wish for the future of *Titanic* exhibitions, it's that the liner itself continues to feature prominently. We're already doing this with the Big Piece, and with deck fittings or furnishings that have stood the test of time, such as the frames for the benches on the promenade decks. We are fortunate to be able to draw on a unique source of documentation: the remarkable photographs taken by Francis Browne, an Irish Jesuit priest and first-class passenger on the Southampton-Queenstown route. During the few hours he spent on board, he photographed the liner and its passengers from every angle; hundreds of images that immortalize the atmosphere and decor of the *Titanic*. It's probably to him we owe the only photograph of the Marconi hall in operation, if it is indeed that of the *Titanic*.

It would be beneficial if future exhibitions devoted more space to archaeological, oceanographic and technical presentations of the work carried out on the wreck and its remains, from deep-sea exploration with manned submersibles or ROV or AUV-type robots, to conservation methods for the artefacts brought up, to historical research around these artefacts—things that RMS Titanic, Inc. has already been doing successfully for years. Some of the salvaged pieces could be integrated into spaces such as the Titanic Belfast Experience, a beautiful space dedicated to the liner, located in the grounds of the shipyard that built it. This Northern Ireland museum currently has no artefacts from the wreck. We could, for example, display a boiler there. The Cité de la Mer in Cherbourg, one of the *Titanic*'s ports of call, is also well suited to exhibitions about the ship. Housed in the former ferry terminal, this museum can accommodate large-scale pieces. For several years now, it has had a permanent exhibition that includes real artefacts, and it evolves with the times.

During a series of dives in 1993 and 1994, we hauled up the *Titanic*'s steam whistles. Curiously, just as the fourth funnel was a dummy, the whistles on the third and fourth funnels were purely decorative. Although identical to the others, they were not powered by steam. Each set of whistles consists of a main whistle and two secondary whistles. The main whistle alone weighs 110 kilos, and the whole unit weighs 360 kilos including connections, all in bronze. At the time, they were the largest whistles in the world. After cleaning, X-raying and compressed-air testing, the components were found to be in excellent condition and declared 'fit for use'. In 1999, prior to the opening of the *Titanic* exhibition in St. Paul-Minneapolis (Minnesota), RMS Titanic, Inc. announced that, for the first time since 1912, *Titanic*'s whistles would be blown again. No fewer than 150,000 people flocked to Saint-Paul for the event. I couldn't make it in person, but I was able to watch it online. When I heard the whistles, live, I got goosebumps, even sitting at home in front of my screen.

As long as the remains of the *Titanic* continue to fire up hundreds of thousands of enthusiasts and as long as the liner continues to live on by sounding her whistles, I will know that we're right to fight this ongoing battle to recover as many artefacts as possible from the wreck before it disappears. They are the liner's historical memory. The *Titanic* will never resurface. But the truth about its sinking will. The Titanic's myths may be alluring, but reality is more powerful than fiction.

Titanic Glossary

Ships:

RMS *Carpathia*: Owned by the British company, Cunard Line, the *Carpathia* was the ship who answered the *Titanic's* call for help. After navigating through an ice field, it arrived two hours after the Titanic sank and the rescued 706 people from lifeboats. The *Carpathia* sank after being torpedoed by a German submarine during WWI.

Mackay-Benett, Minia, Montmagny and *Algerine*: These are the boats chartered by White Line in the days immediately after the *Titanic* went down. They collected approximately 330 bodies before the search was called off on 8th June.

Submersibles and support boats:

Nautilie: A submersible owned by Ifremer, commissioned in 1984. It is 8 metres long and is capable of holding three people. It can reach depths of 6,000 metres. This was the main submersible used to recover items from the *Titanic*.

Robin: The ROV (remotely operated vehicle)/inspection robot used by the *Nautilie*.

Nadir: The support vessel from which the *Nautilie* was launched during the exploration of the *Titanic* wreck.

Alvin: A submersible owned by the US Navy and operated by Woods Hole Oceanography Institution.

Titan: Owned by OceanGate, the *Titan* was the first commercial submersible which reached depths of 4,000 metres. It was also the first crewed submersible with a hull constructed of titanium and carbon fiber. On 18th June 2023 the *Titan* imploded, instantly killing the 5 people on board.

Ocean Voyager: A research vessel which was used as a studio for the production of a documentary for the Discovery Channel, NBC and Canal+.

Abeille Supporter: The support boat that hoisted the Big Piece out of the ocean.

Remora 6000 ROV (remotely operated vehicle): In 2004 seventy-eight artefacts were recovered from the *Titanic* debris field during the ROV's ten dives. The *Remora* was also used during a 2019 expedition part of the Five Deeps Expedition

Organisations:

Ifremer: The French research institution for exploration of the sea. Along with Woods Hole Oceanographic Institution they discovered the wreck of the *Titanic*.

RMS Titanic Inc: The company financed and organized expeditions to the Titanic wreck and recovered over 5500 artefacts, which are now displayed in two permanent immersive exhibitions. It is now part of the Experiential Media Group. RMS Titanic Inc were granted exclusive rights to salvage the wreck by the United States District Court, however these laws have no bearing on citizens who are not from the USA.

Titanic Ventures: The previous name for RMS Titanic Inc, who co-funded the first salvage expedition to the wreck.

Titanic Historical Society: The group, which was established in 1963, opposed the salvage of the wreck and felt it should be left undisturbed.

Woods Hole Oceanographic Institution: A non-profit organization dedicated to ocean research, exploration, and education. Robert Ballard of WHOI discovered the wreck of the Titanic along with Jean-Luis Michel of Ifremer.

National Oceanic and Atmospheric Administration (NOAA): The organization carried out expeditions to the wreck of the Titanic using submersibles, but opposed the salvaging of artefacts.

UNESCO (United Nations Educational, Scientific and Cultural Organization): In 2012 Titanic became world heritage site.

EverGreene Architectural Arts: the company charged with restoring the Big Piece. Treatment consisted of neutralizing chlorides in an alkaline bath; mechanical cleaning; anti-corrosion treatment with tannic acid; and application of microcrystalline for protection and finishing. This treatment took place over two years, from 1998 to 2000.

OceanGate: A privately owned company who had crewed submersibles for tourism, industry, research, and exploration. The company was founded in 2009 by Stockton Rush and Guillermo Söhnlein. Stockton Rush was on the Titan submersible when in it imploded in 2023 and OceanGate have suspended business indefinitely.

THANKS

I would like to thank the *Nautile*, Genavir and Ifremer teams I worked with, and from whom I have learnt a great deal over the past ten years, particularly about the *Titanic*:

Guy Sciarrone, chief engineer and chief pilot of the *Nautile*
Jean-Michel Nivaggioli
Yves Potier
Jean-Paul Justiniano
Max Dubois
Christian Le Guern
Norbert Compagnot
Olivier Cipriani
Georges Arnoux
Jean-Pierre Labbé
Patrick Cheilan
Edmond Tousaint

Jean-Jacques Kaïoun

Robert Tarquini

Frédéric Biguet

Jean-Louis Michel

Charlie Blasi

Frédéric Hennebelle

Henri Martinosi

Gérard Vultaggio

Guy Leclere

Jean-Pierre Chopin

Pierre-Yves Le Bigot

Éric Lacoupelle

André Bonfiglio

Pierre Valdy

Yann Keranflec'h

Yves Houard

Xavier Placaud

Alexis Peuch

Bernard Leduc

Sylvain Pascaud

Pierre Triger

As well as all those I've forgotten, to whom I apologise.

I'd also like to thank all the Aqua+ team who took part in two expeditions aboard the *Ocean Voyager*, including my friends Christian Pétron and Marius Orsi, and my son Julien Nargeolet.

My thanks also to Jean-Noël Mouret, who did a considerable amount of writing and research, and Hélène Vaveau-Schmitt, who had the patience to read and re-read the text, and to suggest simplifications and adaptations. This book is the fruit of a team effort.

Thanks to all of you.